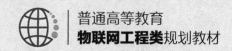
普通高等教育
物联网工程类规划教材

物联网
专业英语

魏旻◎编著

人民邮电出版社
北京

图书在版编目（CIP）数据

物联网专业英语 / 魏旻编著. -- 北京：人民邮电出版社，2017.1
普通高等教育物联网工程类规划教材
ISBN 978-7-115-43874-4

Ⅰ. ①物… Ⅱ. ①魏… Ⅲ. ①互联网络－英语－高等学校－教材②智能技术－英语－高等学校－教材 Ⅳ. ①TP393.4②TP18

中国版本图书馆CIP数据核字(2016)第254268号

内　容　提　要

本书是物联网专业英语教材，针对高等院校物联网工程类专业的需求而编写，选材广泛，体系完整，内容翔实。书中首先介绍了物联网的概貌、发展历程及参考架构，然后重点讲解了RFID、传感器、Wi-Fi、ZigBee、6LoWPAN、3G、5G、LTE、物联网安全、云计算等关键技术和相关应用，最后通过智能电网和智能家居等领域的典型案例对上述技术进行了进一步解读和说明。每篇课文后均配有生词、短语、缩略语及习题，用以巩固读者对课文的理解；每章后附有部分参考译文及练习题参考答案，并根据情况设置科技文献阅读、翻译、写作技巧、物联网信息检索、物联网国际标准化情况等内容，可供教师及学生学习参考。

本书既可作为高等本科院校、高等职业院校物联网相关专业的专业英语教材，也可作为从业人员自学的参考书。

◆ 编　著　魏　旻
　　责任编辑　税梦玲
　　责任印制　沈　蓉　彭志环

◆ 人民邮电出版社出版发行　北京市丰台区成寿寺路11号
　　邮编　100164　电子邮件　315@ptpress.com.cn
　　网址　http://www.ptpress.com.cn
　　北京九州迅驰传媒文化有限公司印刷

◆ 开本：787×1092　1/16
　　印张：15.5　　　　　　　　　2017年1月第1版
　　字数：402千字　　　　　　　2024年9月北京第8次印刷

定价：49.80元

读者服务热线：(010)81055256　印装质量热线：(010)81055316
反盗版热线：(010)81055315

前　言

物联网技术处于高速发展之中，已经成为各国研究的热点，其国际合作化特征尤为明显。从业人员只有提高专业英语水平，才能便于获得最新、最先进的专业知识，因此，学好物联网专业英语是物联网领域从业人员开展科学研究、国际合作和工程实践的重要基础和保障。

本书作者所在的国家工业物联网示范性国际科技合作基地——工业物联网与网络化控制教育部重点实验室，在物联网标准方面与国际同步，在物联网技术方面处于国际前沿，其核心技术正在形成专利保护群，相关研究成果也得到了国际上的广泛认可。本书作者长期从事物联网技术方面的研究，牵头负责物联网国际标准的制定和相关科研项目（包括国际合作项目）的研究，为本书的编写奠定了坚实基础。在本书的编写过程中，作者参考了近年国内外出版的多本同类教材，并结合了作者多年来在物联网相关技术及"物联网专业英语"课程教学上的经验，形成本书如下主要特点。

1. 立足于物联网工程培养方案和物联网工程专业英语教学大纲，选取物联网专业的学生最需要掌握的关键技术专业英语作为教学内容。

2. 融入科技英语翻译技巧和科技文章撰写方法等内容。针对学生在科技文献翻译和撰写中存在的疑问，给出科技文献翻译和撰写的技巧，提高学生的专业英语阅读和写作能力。

3. 选材强调新颖、准确。本书参考了国外期刊中关于物联网技术的最新热点文章，强调物联网专业英语在技术上的准确性、基础性和学术性，可帮助学生扩展知识面，使他们更深入地理解物联网技术的发展状况。

4. 强调两个维度，即，第一维度基础、理论和学术，第二维度产业、案例和应用。在关键技术基础理论讲解的基础上，本书引入创新性的物联网技术实用模式内容，在其后紧跟与这些关键技术相结合的实际应用案例，不再将关键技术与应用分开讨论，让学生在学习中能够将理论与实践结合起来。

5. 突出技术并结合最新科研成果。本书依托重庆邮电大学的信息技术特色优势，结合国家工业物联网国际科技合作基地、工业物联网与网络化控制教育部重点实验室项目、重庆市物联网工程技术研究中心团队的最新成果，满足国家战略、行业发展及地方经济对信息类人才的需求。

6. 本书融入了作者多年来参加ISO/IEC JTC1 WG10 物联网标准工作组、国家重大科技项目、国家863项目研究和参与制定国际标准的研究成果，可让读者对物联网体系架构、关键技术、标准与系统开发等有系统、全面、深入的理解。

7. 单元设置特色鲜明，每一单元包含以下内容：

- 课文——选材广泛、风格多样、切合实际的两篇英语专业文章；
- 单词——课文中出现的新词；
- 词组——课文中的常用词组；
- 缩略语——课文中出现的、业内人士必须掌握的缩略语；
- 习题翻译——培养读者的翻译能力；
- 参考译文——让读者对照理解，从而提高翻译能力；
- 科技文献的阅读、翻译、写作技巧及物联网文献检索等内容——提升读者物联网科技文献阅读、写作能力。

本书由魏旻组稿、统稿、编著，岳眘和陈俊华高级工程师、刘琳博士、王江博士参与了部分章节的整理和英文的校对，研究生毛久超、李潇伶、张琼、杨涛、庞巧、庄园参与了各章节的一些编写和英文的翻译，承蒙曲阜师范大学岳守国教授通读全书，提出修改意见。另外，在本书的编写过程中，王平教授给予了很多非常宝贵的意见，在此特别表示感谢。

编　者

2016年8月

目 录

Unit 1

Unit 1　Introduction to IoT　　　　　　　1
  Passage A　An Overview of IoT　　　　2
  Passage B　Smart Objects and Related Technology　9
  Passage C　IoT Endpoint Monitoring Systems　16
  Exercises　　　　　　　　　　　　　18
  参考译文　Passage A　物联网概述　　19
  参考答案　　　　　　　　　　　　　24
  专业英语阅读策略与技巧　　　　　　25

Unit 2

Unit 2　Architectures of IoT　　　　　　26
  Passage A　IoT Reference Architectures　27
  Passage B　Sensor Network Reference
　　　　　　Architecture　　　　　　　33
  Passage C　Industrial Internet Reference
　　　　　　Architecture　　　　　　　36
  Exercises　　　　　　　　　　　　　41
  参考译文　Passage A　物联网参考架构　42
  参考答案　　　　　　　　　　　　　46
  科技翻译的标准　　　　　　　　　　47

Unit 3

Unit 3　RFID	**48**
Passage A　An Overview of RFID	49
Passage B　RFID Applications	54
Passage C　RFID for Medical Application and Pallet Tracking	57
Exercises	62
参考译文　Passage A　RFID概述	63
参考答案	66
科技文翻译技巧一：分句与合句	67

Unit 4

Unit 4　Sensor	**68**
Passage A　Sensor Characteristics and Measurement Systems	69
Passage B　Capacitive Sensor Working Principle	74
Passage C　Sensor Networks Application in Agriculture and Forest Fire Prevention	76
Exercises	80
Passage A　传感器特性及测量系统	81
参考答案	83
科技文翻译技巧二：增补与省略	84

Unit 5

Unit 5　Wi-Fi	**85**
Passage A　An Overview of Wi-Fi	86
Passage B　CSMA/CA Mechanism	91
Passage C　Surveillance System Applications Over Public Transportation	94
Exercises	97
参考译文　Passage A　Wi-Fi技术概述	98
参考答案	101
科技文翻译技巧三：转性与变态	102

Unit 6 Wireless Sensor Networks   103

- Passage A   ZigBee Technology   104
- Passage B   IEEE 802.15.4 Technology and 6LoWPAN   109
- Passage C   Online Monitoring System for Transmission Lines and Sensing of Traffic Flows   116
- Exercises   119
- 参考译文   Passage A   ZigBee技术   120
- 参考答案   123
- 英文科技论文写作   124

Unit 7 Mobile Communication Technology 128

- Passage A   3G & 4G Technology   129
- Passage B   Modulation (FSK, GFSK)   132
- Passage C   5G and LTE-M   136
- Exercises   143
- 参考译文   Passage A   3G&4G技术   144
- 参考答案   146
- 如何用英语撰写科技论文   147

Unit 8 IoT Security   149

- Passage A   An Overview of IoT Security   150
- Passage B   Access Control   155
- Passage C   SDN Security Considerations in the Data Center   159
- Exercises   163
- 参考译文   Passage A   物联网安全   164
- 参考答案   168
- 物联网专业信息检索   169

Unit 9

Unit 9	**Cloud Computing**	**170**
Passage A	An Overview of Cloud Computing	171
Passage B	Mobile Cloud Computing and Granules Runtime for Cloud Computing	175
Exercises		178
参考译文	Passage A 云计算技术	179
参考答案		181
物联网相关国际学术期刊		182

Unit 10

Unit 10	**IoT for Smart Grid**	**186**
Passage A	The Introduction of Smart Grid	187
Passage B	An IoT-based Energy-management Platform for Industrial Facilities	193
Passage C	Smart Grid Applications	211
Exercises		214
参考译文	Passage A 智能电网介绍	215
参考答案		219
物联网国际标准组织		220

Unit 11

Unit 11	**Smart Home**	**222**
Passage A	Smart Home Overview	223
Passage B	EnOcean Wireless Technology for Smart Home	227
Passage C	Smart Home System for Elderly	231
Exercises		233
参考译文	Passage A 智能家居概述	234
参考答案		236
英文简历写作技巧		237

Unit 1

Introduction to IoT

Passage A   An Overview of IoT

Passage B   Smart Objects and Related Technology

Passage C   IoT Endpoint Monitoring Systems

Passage A   An Overview of IoT

I. An Introduction

Today sensors appear everywhere. We find that there are sensors in our vehicles, smart phones, factories controlling CO_2 emissions, and even in the ground monitoring soil conditions in vineyards. While it seems that sensors have been used for quite a while, the research on wireless sensor networks (WSNs) can also date back to 1980s, and it is only in 2001 that WSNs began to attract an increased interest from industrial and research perspectives. This is due to the availability of inexpensive, low powered miniature components like processors, radios and sensors that were often integrated on a single chip (system on a chip (SoC)).

The idea of internet of things (IoT) was developed in parallel to WSNs. The term of internet of things was proposed by Kevin Ashton in 1999 and refers to uniquely identifiable objects and their virtual representations in an "internet-like" structure.

Various outlooks exist for defining the significant opportunity for globally interconnected and networked "smart things" resulting in an Internet of Things. The following statistics demonstrate that while the estimated volume of connected things may vary, the market impacts are projected to be quite significant.

- Global machine-to-machine connections will rise from two billion at the end of 2011 to 12 billion at the end of 2020. (Machina Research)
- The Internet of Everything — connected products ranging from cars to household goods — could be a $19 trillion opportunity. (Cisco Systems Inc. Chief Executive Officer John Chambers)
- Only 0.6% of physical objects that maybe part of Internet of Things are currently connected. (Cisco)
- The vision of more than 50 billion connected devices will see profound changes in the way people, businesses and society interact. (Ericsson)
- The Internet of Everything could boost global corporate profits by 21 percent by 2022, Cisco said. By 2020, 50 billion objects will be connected to the Internet, according to the slides. (Bloomberg)

While IoT does not assume a specific communication technology, wireless communication technologies will play a major role, and in particular, WSNs will facilitate many applications and many industries. The small, rugged, inexpensive and low powered WSN sensors will bring the IoT to even the smallest objects installed in any kind of environment, at reasonable costs. Integration of these objects into IoT will be a major evolution of WSNs. A WSN can generally be described as a network of nodes that cooperatively sense and control the environment, enabling interaction between persons or computers and the surrounding environment. In fact, the activity of sensing, processing and communication with a limited amount of energy ignites a cross layer design approach, typically requiring the joint consideration of distributed signal/data processing, medium access control, and communication protocols.

Through synthesizing existing WSNs applications as part of the infrastructure system, potential new applications can be identified and developed to meet future technology and market trends. For instance, WSN technology applications for smart grid, smart water, intelligent transportation systems, and smart home generate huge amounts of data, and this data can serve many purposes.

Additionally, as the modern world shifts to this new age of WSNs in the IoT, there will be a number of legal implications that will have to be clarified over time. One of the most pressing issues is the ownership and use of the data that is collected, consolidated, correlated and mined for additional value. Data brokers will have a flourishing business as the pooling of information from various sources will lead to new and unknown business opportunities and potential legal liabilities. The recent US National Security Administration scandal and other indignities have shown that there is wide interest in gathering data for varied uses.

One of the more complex issues which arise within this new world is the thought of machines making autonomous decisions, with unknown impacts on the environment or society within which it functions. This can be as simple as a refrigerator requesting replenishment for milk and butter at the local store for its owner, or as complex as a robot that has been programmed to survive in a harsh environment that originally did not foresee human interaction. It can also be as simple as a vehicle that records its usage, as does the black box in the aerospace industry, but then not only providing the information to understand the cause of an accident, but also is using evidence against the owner and operator. For example, a machine will notifiy legal authorities if it is used against the law.

It comes to the point where a machine starts acting as if it were a legal entity. The question of liability starts to get fuzzy and the liability for the "owner" and "operator" of the machine gets more difficult to articulate if there is no real human intervention in the actions of the machine or robot.

This is certainly the worst case scenario, but the question is how to balance the cost of potential liabilities with the benefits of IoT solutions? This quickly starts to become more of a societal or ethical, and moral discussion. That is what we usually refer to as generational shifts in values –but the IoT trend will not wait for a generation.

II. Definitions and Market Requirements of IoT

ISO/IEC JTC 1/SWG 5 (Special Group 5 - IoT) gave IoT a definition as follows:

"An infrastructure of interconnected objects, people, systems and information resources together with intelligent services to allow them to process information of the physical and the virtual world and react".

To ensure that standards support the anticipated size of the IoT, ISO/IEC JTC 1/SWG 5 (Special Group 5 - IoT) has determined that the following issues and topics need to be considered as the market requirements of the IoT:

- Ease of use
- Data Management
- Security
- Privacy/Confidentiality
- Regulation
- Infrastructure
- Awareness of service
- Accessibility and usage context
- Cohesive set of standards across all standards domains
- Distributed IT and Communications Management – e.g. software defined structures and virtualized systems management (e.g. SDN / NFV)
- Cross domain / vertical routing management (e.g. one to many distribution flows across the

applications domains)
- Governance of IoT

III. Standards of IoT

Standardization is a major prerequisite to achieve interoperability, not only between products of different vendors, but also between different solutions, applications and domains. The latter is of special interest to IoT and WSN as common access to devices, sensors and actors from various application domains leading to new cross domain applications is the major concern of IoT.

Interoperability has to be considered at different layers ranging from component to communication, information, function and business layer. The component layer basically reflects not merely devices like sensors and actuators, but also gateways and servers which run the applications. The communication layer is responsible for the data exchange between the components while the information layer represents the actual data. The function layer is concerned with the functionality which can be not only software applications, but also hardware solutions. At the business layer the business interactions are described. From the WSN and IoT approach to provide information exchange between "things" and applications covering various application domains, common communication and information layer standards are of main interests, but generic functions might also be used by different application areas. At the component layer we will find various types of devices, but still standards defining for example form factors and connectors for modules (e.g. wireless modules, control processing unit (CPU) boards) can make sense.

As a prerequisite for the successful standardization, use cases and requirements have to be collected and architecture standards are needed to structure the overall system and identify the relevant functions, information flows and interfaces.

As WSN will be used in the wider context of IoT, IoT standards and standardization activities are also considered. This concerns particularly the higher communication protocol, information and function layer.

IEEE 802.15.4 is the most relevant communication standard for the WSN. It defines the physical and link layer for short-range wireless transmission with low power consumption, low complexity and low cost. It uses the ISM frequency bands at 800/900 MHz and 2.4 GHz. IEEE 802.15.4 is the foundation for other standards like ZigBee, WirelessHART, WIA-PA and ISA.100.11a, which defines regional or market specific versions. The base standard was published in 2003 and revised in 2006 and 2011. Various amendments have been added to cover additional physical layer protocols, regional frequency bands and specific application areas. Current work is covering additional frequency bands (e.g. TV white space, regional bands), ultralow power operation and specific applications like train control.

Bluetooth is also a wireless short range protocol defined by the Bluetooth Special Interest Group. With Bluetooth 4.0 they have included a low energy protocol variant for low power applications. RFID is not only used in the WSN context, but is of general interest to IoT.

ISO/IEC JTC 1/SC 31 is one of the major standardization drivers with its ISO/IEC 18000 series of standards defining diverse RFID technologies. Other bodies like ISO, EPCglobal and DASH7 have either contributed to or used these standards.

While the lower communication layers are often specific for a certain application approach like WSN, the network and higher communication layer should preferably use common protocols in order to allow interoperability across networks. Still specific requirements of certain technologies, such as low power consumption and small computational footprints in the case of WSNs, have to be taken into account. The IP protocol suite is today the defacto standard for these layers.

While previous domain specific standards have defined their own protocol stack they all move today to IP. It is the preferred solution in the case of WSN and IoT IPv6. The IPv6 standards set (network to application layer) from Internet Engineering Task Force (IETF) (RFC 2460 and others) is available and stable. In order to support low power constrained devices and networks, especially considering IEEE 802.15.4, IETF is working on specific extensions and protocols. The 6LoWPAN working group defines the mapping of IPv6 on IEEE 802.15.4 (e.g. RFC 6282). The roll working group considers routing over low power and lossy networks (e.g. RFC 6550). The constrained application protocol (CoAP) working group defines an application protocol for constrained devices and networks. This is an alternative to the HTTP protocol used for RESTful web services taking into account the special requirements of constrained devices and networks.

The ZigBee specifications enhance the IEEE 802.15.4 standard by adding network and security layers and an application framework. They cover various application areas like home and building automation, health care, energy and light management and telecom services. The original Zigbee specifications define their own network and application layer protocols, while the latest Zigbee IP specification builds on IPv6 and CoAP.

For the actual data exchange between applications various approaches exist, often using a service oriented architecture (SOA). Examples are OPCUA which is an IEC standard and SOAP, WSDL and REST defined by World Wide Web Consortium (W3C). XML as defined by W3C is the commonly used encoding format. In the context of WSN it has to be considered how far these protocols fit to constrained devices and networks.

The Open Geospatial Consortium (OGC) has defined a set of open standards for integration, interoperability and exploitation of web-connected sensors and sensor-based systems (sensor web enablement).

For the management of devices and networks the SNMP protocol defined by IETF is widely used. NETCONF is a new approach for network management in IETF. Currently activities have started to cover management of constrained devices and networks explicitly in IETF. Other devices management protocols considered for IoT are TR-69 from Broadband Forum (BBF) and Open Mobile Alliance (OMA) Device Management. Semantic representation of the information is an important issue in WSN and IoT in order to ease knowledge sharing and auto-configuration of systems and applications. W3C is defining the base protocols like RDF, RDFS and OWL in its semantic web activities. Again the specific requirements of constrained networks and devices have to be taken into account. Furthermore semantic sensor network ontology has been defined. For querying geographically distributed information OGC has defined GeoSPARQL. The European Telecommunications Standards Institute (ETSI) TC SmartM2M has started from use cases and requirements for several application areas to develop M2M communication architecture and the related interfaces between devices, gateways, network notes and applications with a focus on offering M2M services. This work is introduced into OneM2M.

ISO/IEC JTC 1/WG 7 (Sensor Networks) has developed the ISO/IEC 29182 services for a sensor network reference architecture and services and interfaces for collaborative information processing. They are working on sensor network interfaces for generic applications and smart grid systems.

ITU has set up a M2M focus group to study the IoT standardization landscape and identify common requirements. Its initial focus is on the health sector. A joint coordination activity (JCA-IoT) shall coordinate the ITU-T work on IoT, including network aspects of identification functionality and ubiquitous sensor networks (USNs). In addition ITU has varies more or less related activities for example on next generation networks including USN, security and identification (naming and numbering).

IEEE has in addition to the 802.15.4 also activities on smart transducers (1451 series) and for ubiquitous green community control (1888 series). Information models, sometimes with semantic representation and even ontology are already available for different application areas like for smart grid from IEC TC 57, industry automation from IEC TC 65 and ISO TC 184 and building automation from ISO TC 205 and ISO/IEC JTC 1/SC 25.

Important in the IoT context are also product data standards as defined for example by IEC SC 3D, identification standards as defined by ISO and ITU and location standards as defined for example by ISO/IEC JTC 1/SC 31 and OGC.

IV. IoT Applications

1. WSN Application in Intelligent Transportation

Wireless sensing in intelligent transportation differs on several points from the traditional concepts and design requirements for WSN. In most cases, sensors can rely on some sort of infrastructure for power supply, for example the aspect of energy efficiency is usually of secondary importance in these systems. WSN applications in intelligent transportation can be subdivided into two categories:

(1) Stationary sensor networks, either on board of a vehicle or as part of a traffic infrastructure.

(2) Floating sensor networks, in which individual vehicles or other mobile entities act as the sensors.

The latter category comprises applications related to the tracking and optimization of the flow of goods, vehicles and people, whereas the former comprises mainly applications that were formerly covered by wired sensors.

Intelligent traffic management solutions rely on the accurate measurement and reliable prediction of traffic flows within a city. This includes not only an estimation of the density of cars on a given street or the number of passengers inside a given bus or train but also the analysis of the origins and destinations of the vehicles and passengers. Monitoring the traffic situation on a street or intersection can be achieved by means of traditional wired sensors, such as cameras, inductive loops, etc. While wireless technology can be beneficial in reducing deployment costs of such sensors, it does not directly affect the accuracy or usefulness of the measurement results.

However, by broadening the definition of the term "sensor" and making use of wireless technology readily available in many vehicles and smart phones, the vehicles themselves as well as the passengers using the public transportation systems can become "sensors" for the accurate measurement of traffic flows within a city.

City logistics is another use case in this area. Urbanization is posing a lot of challenges, especially in rapidly developing countries where already huge cities are still growing and the increasingly wealthy population leads to a constantly rising flow of goods into and out of the city centers.

Delivery vehicles account for a large portion of the air pollution in the cities, and streamlining the flow of goods between the city and its surroundings is the key to solving a lot of the traffic problems and improving the air quality. A promising approach towards reducing the traffic load caused by delivery vehicles is the introduction of urban consolidation centers (UCCs), i.e. warehouses just outside the city where all the goods destined for retailers in a city are first consolidated and then shipped with an optimized routing, making the best possible use of truck capacity and reducing the total number of vehicles needed and the total distance travelled for delivering all goods to their destinations.

To achieve such optimization, careful analysis and planning of traffic flows in the city as well as monitoring of the actual flow of the goods are needed. The challenges and the solutions are similar to the ones discussed above, but with a finer granularity. Rather than just tracking a subset of vehicles as they move through the city, tracking of goods at least at a pallet level is required. The pallet (or other packaging unit) thus becomes the "sensor" for measuring the flow of goods, and a combination of multiple wireless technologies (GPS, RFID, WLAN, cellular) in combination with sophisticated data analysis techniques are applied to obtain the required data for optimizing the scheduling and routing of the deliveries and ensure timely arrival while minimizing the environmental impact of the transportation.

2. IoT Applications in Smart Grid

The power grid is not only an important part of the electric power industry, but also an important part of a country's sustainability. With the dependence on electric power gradually increasing, demand for the reliability and quality of the power grid is also increasing in the world. Utilities, research institutions and scholars have researched how to modernize the power grid to one that is efficient, clean, safe, reliable, and interactive. A smart electricity grid opens the door to new applications with far-reaching impacts: providing the capacity to safely integrate more renewable energy sources (RES), electric vehicles and distributed generators into the network; delivering power more efficiently and reliably through demand response and comprehensive control and monitoring capabilities; using automatic grid reconfiguration to prevent or restore outages (self-healing capabilities); enabling consumers to have greater control over their electricity consumption and to actively participate in the electricity market. General architecture transmission line based on WSNs as shown in Figure 1.1.

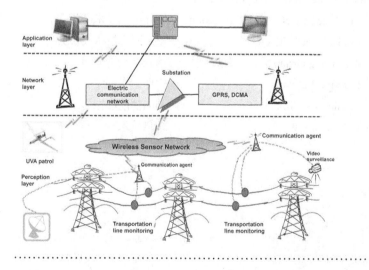

Figure 1.1　General architecture transmission line based on WSNs

Sensors will be a key enabler for the smart grid to reach its potential. The idea behind the "smart" grid is that the grid will respond to real time demand; in order to do this, it will require sensors to provide this "real time" information. WSNs as "smart sensing peripheral information" can be an important means to promote smart grid technology development. WSN technology in the smart grid will also further promote the industrial development of WSNs.

Ⅴ. New Words

sensor ['sensə]	n.	传感器
integrated ['ɪntɪgreɪtɪd]	adj.	综合的；完整的；互相协调的
object ['ɒbdʒɪkt]	n.	目标；对象；客体
proliferate [prə'lɪfəreɪt]	vi.	增殖；扩散；激增
process ['prɑsɛs]	v.	加工；处理
synthesize ['sɪnθəsaɪz]	v.	合成；不同元素间的整合
cohesive [kəʊ'hiːsɪv]	adj.	有结合力的；紧密结合的；有黏着力的

Ⅵ. Phrases

internet-like	网络化
smart grid	智能电网
smart water	智能水资源处理
smart home	智能家电
data Management	数据管理
awareness of service	服务意识

Ⅶ. Abbreviations

WSNs Wireless Sensor Networks	无线传感器网络
SoC System on a Chip	片上系统
IoT internet of things	物联网
ISO International Standardization Organization	国际标准化组织
IEC International Electro technical Commission	国际电工技术委员会
JTC1 Joint Technical Committee 1	第一联合技术委员会
SWG5 Special Working Group 5	第五特别工作组
SDN Software Define Network	软件定义网络
NFV Network Functions Virtualization	网络功能虚拟化
IT Information Technology	信息技术

Passage B Smart Objects and Related Technology

Smart objects represent the middle ground between computing and telephony, borrowing from both. From its computing heritage, smart objects have assumed the culture of engineering evolvable systems. This is important because at this point, it is impossible to fully specify the expected behavior of future smart object systems, even if we have a good idea of where smart objects are heading today. From its telephony heritage, smart objects have applied the principles from connecting disparate systems that may be managed by different companies and organizations. Smart objects are not manufactured by a single organization, but by multitudes of different people and parties. Smart object technology must be both evolvable and standardized.

We discuss today's smart objects as shown in Figure 1.2: embedded systems, ubiquitous and pervasive computing, mobile telephony, telemetry, wireless sensor networks, mobile computing, and computer networking. Some of these areas come from the computing heritage and some from the telephony heritage. Some have sprung out of academic research communities, some from an industrial background. What they have in common, however, is that they either deal with computationally assisted connectivity among physical items, wireless communication, or deal with interaction between the virtual and the physical world.

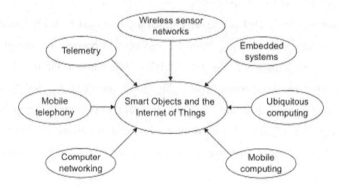

Figure 1.2 Smart objects are the intersection of embedded systems, ubiquitous computing, mobile telephony, telemetry, wireless sensor networks, mobile computing, and computer networking.

Ⅰ. Embedded Systems

An embedded system is a computer embedded in something other than a computer. Under this definition, any system that has a microprocessor is an embedded system with the exception of PCs, laptops, and other equipment readily identified as a computer. Thus this definition of an embedded system would include smart objects. Figure 1.3 illustrates different types of embedded systems.

The primary difference between a traditional embedded system and a smart object is that communication is typically not considered a central function for embedded systems, whereas communication is a defining characteristic for smart objects. Although there are many examples of communicating embedded systems, such as car engines with embedded microprocessors that can communicate their status information with a computer connected to the engine at service time, these systems are not defined by their ability to communicate. A car engine that cannot communicate can still operate as a car engine. In contrast, a smart object such as a wireless temperature sensor deprived of its communication abilities would no longer be able to fulfill its purpose.

Figure 1.3 Embedded Systems

II. Ubiquitous and Pervasive Computing

Ubiquitous computing, also called pervasive computing, is a field of study based on the concept of what happens when computers move away from the desktop and become immersed in the surrounding environment. Ubiquitous computing, as a research discipline, originated in the mid-1980s. Mark Weiser, a professor at MIT, published two short notes titled "Ubiquitous computing #1" and "Ubiquitous computing #2.", which described that computing would move into our daily environment, living in "the woodwork of everywhere" as exemplified in Figure 1.4. He criticized the trend of making computers exciting objects in their own right. He took a different perspective: instead of making computers the central object, they would become invisible.

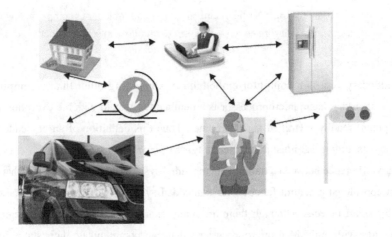

Figure 1.4 Ubiquitous computing is a vision for the future of computers where computing moves into everyday objects.

Ubiquitous computing has become an established academic research field with several major annual conferences and a number of scientific journals. Hundreds of doctoral theses have been written about this topic over the last two decades. As an academic discipline, ubiquitous computing places a strong focus on building real systems that embody its ideas. There is a long string of important prototype systems that come from the ubiquitous computing community. These prototypes have been instrumental in pursuing the field of ubiquitous computing as well as demonstrating the feasibility of an ever-connected world. Wearable computing is a field that has grown out of the ubiquitous computing community. With wearable computing, the computing infrastructure moves onto the body of its users or into their clothing. Wearable computers make ubiquitous computing truly person-centric.

Smart objects owe much of their history to ubiquitous computing. Many of the early developments and vision in ubiquitous computing directly apply to smart objects. Whereas ubiquitous computing is interested in the interaction between ubiquitous computing systems and humans, the area of smart objects takes a more technical approach. Much of the technology developed for smart objects has a direct applicability to ubiquitous computing. Similarly, most of the designs that have been developed within the ubiquitous computing community can be applied to smart objects as well.

III. Mobile Telephony

Mobile telephony grew out of the telephony industry with the promise of ubiquitous access to telephony. Today, mobile telephony not only provides telephony everywhere, but also Internet access. In the late 1990s, nearly 20% of the population in the developed world had a mobile telephone. In 2008, there were more than 4 billion mobile telephony subscribers.

Mobile telephony is often called cellular telephony, and mobile phones are called cell phones, because of the structure of the wireless networks in which mobile phones operate. The network is divided into cells where each phone is connected to exactly one cell at any given time. A cell covers a physical area whose size is determined by the network operator. Since each cell typically handles a limited number of simultaneous phone calls, network operators plan their networks so that cells are smaller and more numerous in areas where operators expect more people to make phone calls. Each cell is operated by a cell tower on which a wireless transceiver base station is mounted. The base station maintains a wireless connection to all active phones in its cell. When the user and the phone move to another cell, the base stations perform an exchange called a handover.

Mobile telephony has given rise to long-range wireless networking technology such as Global System for Mobile communications (GSM), General Packet Radio Service (GPRS), Enhanced Data Rates for GSM Evolution (EDGE), and Universal Mobile Telecommunications System (UMTS) as well as short-range wireless communication technology such as Bluetooth (IEEE 802.15.1). Long range communication is used to transmit voice and Internet data from the mobile phone to the nearest base station. Short-range wireless communication is used for communication between the phone and wireless accessories such as wireless headsets.

Mobile telephony has revolutionized the way we think of personal connectivity. Telephony used to be restricted to a few physical locations: we had a phone at the desk in our office and a few phones at strategic locations in our homes, such as the kitchen or next to the TV. As telephony became mobile, we stopped thinking about telephony as location-bound, but as a ubiquitous always-on service, available everywhere. Mobile telephony not only revolutionized person-to-person access, but changed the way we view network access. With modern smart phones, Internet access is no longer confined to PCs; it is truly ubiquitous. With a few quick button presses, e-mail, instant messaging, and the World Wide Web are immediately available. Instant Internet access is equally available in foreign countries, even if it sometimes costs a small fortune.

The way mobile telephony changed the general view on connectivity is an important factor for the continued development of smart objects. As we are now accustomed to thinking of connectivity as ubiquitous, we are equally accustomed to thinking of access to smart objects as ubiquitous.

IV. Telemetry and Machine-to-machine Communication

The word telemetry is a portmanteau of the Greek words tele (remote) and metron (to measure). Telemetry is, as the name implies, about performing remote measurements. Machine-to-machine communication is a generalization of telemetry that implies autonomic communication between nonhuman operated machines and is central to the concept of telemetry. Telemetry is used to transmit information about current temperature, humidity, and wind from distant weather stations (Figure 1.5). Telemetry is used to transmit fuel consumption data from trucks so that the owner can optimize the truck's routes to save on fuel costs, and as a consequence reduce pollution.

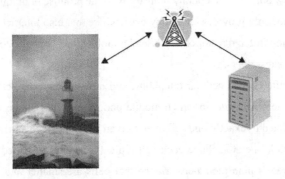

Figure 1.5　Telemetry and Machine-to-machine Communication

The concept of machine-to-machine communication and telemetry is also used in shorter distances. Today's pacemakers (devices that are implanted in the hearts of people who have had a heart attack) frequently include a device called a "telemetry coil." This allows a doctor to monitor the pacemaker's activity without surgery. Instead, the doctor uses a device that creates a low-power electromagnetic field near the patient. The telemetry coil reacts to the electrical field by modulating it creating a low-power communication mechanism with which information can be transferred from the patient's heart to the doctor.

Telemetry and machine-to-machine communication are similar to smart objects because they are both used to perform large-scale measurements. With telemetry, these measurements can be performed from a remote site without direct physical access. Remote access using telemetry is most often performed with existing mobile telephony networks such as GSM or 3G (UMTS), or via dedicated networks such as the Inmarsat satellite network. Smart objects are not only used for measurements and sensing, but also affect their environment by using actuators. Nevertheless, much of the remote access technology developed for telemetry systems can be used with and applied to smart object systems.

V. Wireless Sensor and Ubiquitous Sensor Networks

Wireless sensor networks have evolved from the idea that small wireless sensors can be used to collect information from the physical environment in a large number of situations ranging from wild fire tracking and animal observation to agriculture management and industrial monitoring. Each sensor wirelessly transmits information toward a base station. Sensors help each other to relay the information to the base station. The research field of wireless sensor networks has been very active since the early 2000s with several annual conferences, many journals, and a large number of annual workshops. Wireless sensor networks are sometimes called ubiquitous sensor networks to highlight the ubiquity of the sensors.

Early work in wireless sensor networks envisioned sensor networks to be composed of so-called smart dust. Smart dust would be composed of large numbers of tiny electronic systems with sensing, computation, and communication abilities. It would be spread over an area where a phenomenon, such as humidity or temperature, was to be measured. Because the dust specks would be so small, they could be dispersed using mechanisms such as air flow. The applications of smart dust would initially be used by the military to track the location of enemies, to signal an alarm when intruders were found, or to detect the presence of a vehicle.

The concept of smart dust was, however, too restrictive for most uses. The limited physical size of the dust specks severely limited possible communication mechanisms and the computational capability of the nodes. Instead, many research groups started building hardware prototypes with a larger physical size that were easier to use for experimentation. The research community around wireless sensor networks has developed many important mechanisms, algorithms, and abstractions. Wireless sensor networks are intended to have a long lifetime. Since wireless sensors typically use batteries, having a long lifetime to translate into reducing the power consumption of the individual nodes. Thus, several power-saving mechanisms have been designed, deployed, studied, and evaluated both in simulators and in actual deployments. Many of these have a direct applicability to smart objects.

Wireless sensor networks have further spurred work in standardization for industrial automation and monitoring. Many of the recent standards in wireless industrial networking, such as Wireless HART and ISA100a, have their roots in the wireless sensor networking community.

The concept of wireless sensor networks is similar to that of smart objects, and much of the development in smart objects has occurred in the community around wireless sensor networks. Wireless sensor networks are composed of small nodes, equipped with a wireless communication device, that autonomously configure themselves into networks through which sensor readings can be transported. Smart object networks are less focused on pure data gathering, but are intended for a large number of other tasks including actuation and control. Furthermore, wireless sensor networks are primarily intended to be operated over a wireless radio communications device. In contrast, the concept of smart objects is not tied to any particular communication mechanism, but can run over wired as well as wireless networks.

VI. Mobile Computing

Mobile computing is the field of wireless communication and carry-around computers, such as laptop computers. In some ways the mobile computing field spun out of work initialized within the ubiquitous computing area. Likewise, the early focus on wireless networking led to wireless communication mechanism research. Work on these mechanisms began in the mid-1980s and led up to the standards around wireless local area networks (Wi-Fi) that started forming in the late 1990s. Today, so-called Wi-Fi hot spots at public places such as coffee houses, libraries, and airports are common. Users may connect to the Internet through this wireless network either gratis or for a fee.

In academia, the field of mobile computing also carried over into the research field of Mobile Adhoc NETworks (MANETs). MANET research focuses on networking mechanisms for wireless computers where no network infrastructure exists. In such situations, routing protocols and other network mechanisms must quickly establish an ad hoc network. The network formation is made in a distributed manner where each node that participates in the network must take part in the network's mechanisms such as routing and access control. The MANET community has developed several important routing protocols for these networks such as the standardized AODV and DSR protocols.

Just as with mobile telephony, the use of mobile computing has permeated the understanding that network access is ubiquitous. As Wi-Fi access has become widespread, we now take connectivity for granted anywhere, instantly.

VII. Computer Networking

Computer networking is about connecting computers to allow them to communicate with each other. Computers are connected using networks as shown in Figure 1.6. These networks were initially wired, but with the advent of mobile computing, wireless networks are available.

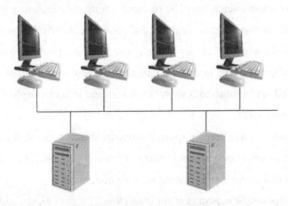

Figure 1.6　Computers are connected using networks

The field of computer networking is significantly older than that of mobile computing. Computer networking began in the early 1960s when the breakthrough concepts of packet-switched networking were first described by Leonard Kleinrock at UCLA. Earlier telephony networks were circuit-switched, and each connection (phone call) created a circuit through the network where all data were transported. With packet-switched networking, no circuits were constructed through the network. Instead, each message was transported as a packet through the network where each node would switch the packet depending on its destination address.

After Kleinrock's breakthrough, ARPANET was created as the first large-scale computer network built on the concepts of a packet-switched network. During the late 1970s and early 1980s, ARPANET was gradually replaced with the early versions of the Internet. ARPANET started to use the IP protocol suite in

1983 before becoming the Internet.

The ARPANET and the Internet were built on a powerful concept called the end-to-end principle of system design, named by an influential paper by Jerome H. Saltzer, David P. Reed, and David D.Clark. The end-to-end principle states that functionality in a system should be placed as long as possible toward the end points. This principle has arguably been one of the most important aspects of the design of the Internet system, because it allowed the system to gracefully support an ever-growing flora of applications from simple e-mail and file transport of the 1980s through the Web revolution of the 1990s transmission to high-speed, real-time video, and audio transmissions of the 2000s. The connection between computer networking and smart objects is evident: communication is one of the defining characteristics of smart objects.

VIII. New Words

heritage ['herɪtɪdʒ]	n.	遗产；传统；继承物；继承权
manufactured [mænjə'fæktʃəd]	adj.	制造的，已制成的
ubiquitous [juː'bɪkwɪtəs]	adj.	泛在的；无所不在的
pervasive [pə'veɪsɪv]	adj.	普遍的；到处渗透的
realtime	adj.	实时的；适时的
synonymous [sɪ'nɒnɪməs]	adj.	同义的；同义词的；等同于……的
rudder ['rʌdə]	n.	船舵；飞机方向舵

IX. Abbreviations

RTOS	Realtime Operating Systems	实时操作系统
GSM	Global System for Mobile communications	全球移动通信系统
GPRS	General Packet Radio Service	通用分组无线服务技术
EDGE	Enhanced Data Rates for GSM Evolution GSM	演进的增强数据率
UMTS	Universal Mobile Telecommunications System	通用移动通信系统
IEEE	Institute of Electrical and Electronics Engineers	电气和电子工程师协会
MANETs	Mobile AdhocNETworks	移动自组织网络

Passage C IoT Endpoint Monitoring Systems

A use case of IoT-based endpoint monitoring systems will be discussed in this section. IoT endpoints (sensors and actuators) are of various types and capabilities distributed where lots of systems are deployed. Many of these endpoints serve single purposes and are located in remote locations. These IoT endpoints are often fragile, low power but serve critical functions in the data or information they gather, transmit or services they provide in the IoT system. IoT endpoints are often too frail to report their own state (power/batterylife, on/off, failure to wake up, performance profile or security profile). This use case proposes the introduction of a capability in the network, local or remote whose sole purpose is monitor or gather state information of sensors/endpoints on a network segment and transmits this information to a central location to enable manage, prevent failures or malfunctioning of critical IoT endpoint, as well as life cycle management of IoT endpoints. The typical node hardware structure is shown as Figure 1.7.

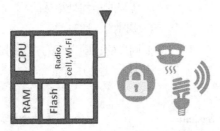

Figure 1.7 The typical node hardware structure

With billions of IoT endpoints and sensors going alive, it will get extremely complicated to monitor and manage the health of an end point sensor, especially if it is constrained, remote and critical to other systems relying on it for information, action. Today, most sensor malfunctions and failures are detected only after the fact in a reactive manner. This might not be acceptable going forward for critical infrastructure and IoT systems that have an impact on human lives. Use case actors and description is shown table 1.1.

Table 1.1 Use Case Actors and Description

Actor Name	Actor Type	Actor Description
IoT endoints (sensors, actuators etc.)		IoT end points work in concert with other IoT endpoints and/or infrastructure.
Endpoint monitoring sensor	Special purpose endpoint (monitor)	Can be configured to monitor required parameters such as patch status, security, failure state etc.
Gateways		Device interconnecting local IoT endpoints to wider communication channels
Telecom network	Network	Interconnects IoT infrastructure
Controller	System	Cloud monitoring platform OR Central monitoring server
Cloud server	Server	Option for monitoring system

Availability of constrained endpoints is a security requirement that is important to critical IoT systems and solutions. IoT endpoints vary widely in capability and they typically work in concert with wider network of IoT endpoints and Infrastructure that can be placed locally, remote and sometimes in inhospitable, not easily accessible remote placements. Status monitoring, life cycle management, upgrades and patch management status can be tedious and difficult. Some embedded systems in sensors are designed to last for up-to 10 years or more and keeping track of their upgrade, patch or battery life status or being aware if they have failed, become defective or are compromised is challenging. When managing thousands of endpoints in the field, if handled manually, can be cumbersome and in some cases impossible. IoT sensor monitoring systems can help alleviate this. Placing a security function in sensor networks whose sole purpose is to monitor sensors within its reach and transmit data continuously, intermittently or periodically subject to the requirement of the IoT solution can help significantly with endpoint lifecycle management. It can help prevent critical safety and security failures, which would otherwise be detected only after the fact and allows for pro-active remediation, repairs, replacement or decommissioning of endpoints.

The health of critical IoT systems will become progressively important, its availability, reliability and security becoming vital, as society progressively moves towards relying on them for automation and critical functions, be it in factories, smart cities, remote field sites or in medical devices. The most fragile component of the IoT system is generally the endpoint or sensor, that is often unprotected, low cost, but where the infrastructure relies on the consistency and availability of the end point and its unfailing performance. This introduces a need for monitoring systems towards the health of the IoT endpoint, sensor or device. Architecture of the endpoint monitoring systems is shown as Figure 1.8.

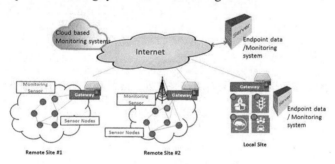

Figure 1.8 Architecture of the endpoint monitoring systems

A monitoring sensor or device placed amidst other functional sensors to gather and keep track of sensor status can function by either periodically detecting the presence of the set of sensor on the network, polling the sensors or tapping into the data/information being sent by the sensors on its designated communication path. The monitoring sensor could be configured with auto-detect capability or by an initially programmed set of data points that the monitoring sensor could update. In addition, maintenance records, in terms of, replacement, software version or latest patch upgrade status may also be recorded. In instances where a sensor fails to wake up, respond or ceases to transmit critical data, the monitoring sensor can generate alarms to central locations. In instances where average battery life is known and tracked the monitoring endpoint could proactively forecast when various sensors are due for a battery refresh thus allowing for proactive maintenance preventing segments of sensor networks from becoming unavailable. This type of sensor monitoring or surveillance would be a preventative security measure to ensure IoT system availability and reliability.

Exercises

Translate the following sentences into Chinese or English.

1. Smartphone apps are likely to provide an important window on the world of IoT, and tablets and other devices will also join in.

2. "The IoT is removing mundane repetitive tasks or creating things that just weren't possible before, enabling more people to do more rewarding tasks and leaving the machines to do the repetitive jobs", says Grant Notman, Head of Sales and Marketing at Wood & Douglas, a company that manufactures wireless communications hardware for IoT application.

3. 两个设备互相的交谈如果不能给所有者带来好处，或提高某种效率，那就没有意义了。

4. 人口增长和日益增加的城市化的"双重打击"意味着物联网的任务就是我们的城市流程化，但在理论上它会帮我们省去很多日常事务。

参考译文 Passage A 物联网概述

Ⅰ. 概述

当前，传感器随处可见。正如我们所知，传感器已被广泛地应用于汽车、智能手机、监控工厂的二氧化碳排放，以及葡萄园地面的土质条件检测。传感器的应用已有一段时间，早在20世纪80年代，无线传感器的研究就开始了。2001年以来，无线传感器网络吸引了工业和学术界的浓厚兴趣，同时，出现了大量的廉价、低功耗微型组件，比如处理器、射频组件和传感器，而且它们经常被集中在一个芯片上（系统芯片SoC）。

物联网的概念与无线传感器网络并行发展。1999年，凯文·阿什顿提出了"物联网"这个词，用来定义物件及其在"网络化"结构中的虚拟表达形式。

由全球化互联和网络"智能"引发的物联网的定义，存在多种观点。以下数据统计表明，虽然连接事物的预估量可能会有所不同，但市场的影响应该极为显著。

➢ 全球机器对机器连接将从2011年的20亿上升到2020年的120亿。（Machina Research）
➢ 万物互联——用互联网将任何产品，从汽车到家用商品连接起来——可能蕴藏着价值190000亿美元的市场。（思科系统公司首席执行官约翰·钱伯斯）
➢ 只有0.6%的实体物件相互连接，这些物件可能会成为物联网的一部分。（思科）
➢ 超过500亿的连接设备将会使人、企业与社会相互作用的方式发生深刻的变化。（爱立信）
➢ 思科在幻灯片里提到，在2022年，万物互联可以提高全球企业利润的21%，在2020年，500亿的物件将会连接互联网。（彭博资讯）

尽管物联网不要求特定的通信技术，但无线通信技术仍将扮演重要的角色，特别是，无线传感器网络将会增加更多的应用和行业。无线传感网中小巧、坚固、价廉、低耗的传感器将会以更合理的成本将物联网引入，并将使其可以被安装在任何环境中的最小物体上。将这些物件融入到物联网将会是无线传感网络的主要发展趋势。无线传感器网络通常被描述为，共同检测和控制环境的网络节点，使人与计算机以及周围的环境相互连接。事实上，检测、处理以及交互活动的能量有限，使用跨层设计方法通常要同时考虑到分布式信号及数据处理、介质访问和通信协议。

通过综合现有的无线传感器网络应用，作为基础设计系统的一部分，潜在的新应用能够被识别和开发出来，以满足未来技术和市场的需要。例如，无线传感器网络技术可以被应用于很多场合，如：智能电网、智能水资源处理、智能交通系统以及智能家居所产生的大量数据，这些数据可以被应用于很多研究之中。

此外，现代社会正在向物联网中的无线传感器网络时代转变，这会带来大量的法律问题，需要在未来得到解决。目前，最迫切需要解决的是数据使用权的问题，以及如何使用收集到的数据来巩固、连接及获得附加价值。数据经纪人将会拥有很好的业务，因为各种来源的信息的汇集会带来新的、未知的商业机会和潜在的法律责任。最近，美国国家安全局的一些丑闻显示，大家对以不同的目的而收集的数据有广泛的兴趣。

在这个全新的世界，一个更复杂的问题是：机器自主决策的思想会对环境和社会造成未知的影响。这就像冰箱可以自主去当地商店为主人补充牛奶和黄油一样复杂，或者像机器人可以通过编程在人类无法存活的恶劣环境中生存。这也像汽车记录其使用率一样简单，像航天工业中的黑匣子，不仅可以用于收集信息了解事故发生的原因，也可以为所有者和操作者提供证据。例如，一台机器发现其自身使用违法便会通知法律当局。

问题的关键在于如果一开始，机器就像一个法律实体。如果没有现实中的人介入操作，责任问题就会变得更模糊，所有者和操作者关于机器的责任问题也更难说清。

这只是最坏的情况，而主要问题在于如何平衡潜在的成本和物联网解决方案所带来的利益。这很快就成为一个关于社会、伦理和道德的讨论。这就是我们经常提到的价值上的年代变化——但是物联网趋势不会等待这一年代的到来。

Ⅱ. 物联网的定义和市场需求

ISO/IEC JTC 1/SWG 5 (IoT)对物联网的定义：

物联网是一个将物体、人、系统和信息资源与智能服务相互连接的基础装置，可以用来处理物理世界和虚拟世界的信息并做出反应。

为确保标准可以支持不同规模的物联网，ISO/IEC JTC 1/SWG 5认为以下问题应该被考虑进物联网的市场需要。

- 易用性
- 数据管理
- 安全性
- 隐私性、机密性
- 监管
- 基础设施
- 服务意识
- 可访问性及使用语境
- 所有标准域中的组内聚标准
- 分布式IT及通信管理（例如，软件定义的结构）
- 虚拟化系统管理（如SDN / NFV）
- 跨域、垂直路由管理（例如，一个到多个分布流跨应用程序域）
- 物联网管理

Ⅲ. 物联网的标准

标准是实现互操作性的重要前提条件，这不仅体现在不同厂商的产品之间，还体现在不同的解决方案、应用和领域之间。后者是物联网和无线传感网特别关注的，因为对设备、传感器和来自不同应用领域的角色所引领的新交叉领域应用是物联网的主要研究内容。

从元件层到通信层、信息层、功能层以及商业层，不同层都需要考虑互操作性。元件层基本反映了传感器、执行器等设备，以及网关和运行应用程序的服务器；通信层负责不同元件间的数据交换；信息层则代表实际的数据；功能层主要关注功能性，包括软件应用和硬件解决方案；商业层描述商业交互。无线传感网和物联网提供了物和不同领域的应用之间信息交换的途径，从这些途径可知，通用的通信和信息层标准，以及用于不同应用领域的通用功能层标准是主要的热点。在元件层，虽然可以找到不同类型的设备，但是用于定义模块构成要素的示例和连接器的标准（如无线模块，控制处理单元）仍然是有意义的。

作为成功标准化的首要条件，不仅需要收集需求和用例，而且需要架构标准来构造整个系统并识别相关的功能、信息流和界面。

随着无线传感网被更广泛地应用于物联网环境中，更多的物联网标准和标准化活动将会涉及其中，尤其涉及更高的通信协议、信息和功能层。

IEEE 802.15.4是无线传感网最相关的通信标准，它定义了低功耗、低复杂度和低开销的短距离无线传输通信协议的物理层和链路层，使用ISM上的800/900 MHz 和 2.4 GHz频段。IEEE 802.15.4为Zigbee、WirelessHART、WIA-PA 和 ISA.100.11a等其他标准奠定了基础，这些标准定义了地区的和市场的特定版本。基本标准发布于2003年，在2006年和2011年发布了修订版。为了涉及其他的物理层协议、地区的频带和特定的应用领域，不同的修订版也相继增加。当前版本涉及附加频带（如电视白色区域、地区频带）、超低功耗操作以及如火车控制等特殊应用。

蓝牙也是一种无线短距离协议，它是由蓝牙特别工作组定义的。蓝牙4.0包含用于低功耗应用的低能耗协议。无线射频识别（RFID）不仅用于无线传感网的环境，也用于物联网。

ISO/IEC JTC 1/SC 31是主要的标准化制定组织之一，它的ISO/IEC 18000系列标准定义了不同的RFID技术。其他如ISO，EPCglobal和DASH7等或是促进或是使用了这些标准。

对于特定的应用方法如无线传感网，较低的通信层都是特定的，而网络层或更高的通信层应该更好地使用通用协议，从而能够支持交叉网络的互操作性。一些技术的特定需求仍需考虑，如低功耗技术、无线传感网中的低计算开销等。IP协议簇是当前这些层事实的标准。

在之前，各领域有特定的标准定义它们自身的协议栈，而现在它们全都转向IP协议。针对无线传感网和物联网，IPv6是首选的解决方案。IPv6标准（RFC2460等）是由IETF制定的，它可用而且稳定。为了能够支持资源受限的设备和网络，尤其考虑到IEEE 802.15.4，IETF正致力于特殊的扩展协议，6LoWPAN工作组定义IPv6在IEEE 802.15.4上的映射（如RFC6282）。Roll工作组考虑低功耗的路由和有损耗的网络（如RFC6550）。CoAP工作组针对受限设备和网络定义应用层协议。这是Web服务的HTTP协议的一个备选方案，它特别地考虑了受限设备和网络的需求。

通过增加网络层安全和应用层架构，ZigBee规范使IEEE 802.15.4标准增强，并覆盖不同的应用领域，如家居、楼宇自动化、医疗保健、能源管理和远程服务等。先前的ZigBee规范定义它们自身的网络层和应用层协议，而最新的ZigBee IP规范是建立在IPv6和CoAP之上的。

应用之间真实数据的交换可借助于各种各样的途径，通常是使用服务向导性的架构（SOA）。如IEC标准OPCUA、SOAP以及W3C定义的WSDL和REST。W3C定义的XML是常用的编码格式。在无线传感网的环境下，必须考虑这些协议如何适用于受限设备和网络。

开放地理空间联盟（OGC）定义了一组开放标准，以使互联网传感器和基于传感器的系统具有集成性、互操作性和开发性。

IETF定义的SNMP协议广泛用于设备和网络的管理。NETCONF是IETF的一种新的网络管理方法。IETF当前的活动明确地涉及受限设备和网络的管理。其他用于物联网的设备管理协议包括BBF的TR-69和OMA设备管理。为了系统和应用之间的知识共享以及自动配置，信息的语义表述是无线传感网和物联网的重要问题。W3C在它的语义网络活动中定义了基本协议，如RDF、RDFS、OWL等。而且，受限设备和网络的特殊要求不得不再次被纳入考虑，此外还需要定义语义传感网本体论。为了查询地理分布信息，OGC定义了GeoSPARQL。欧洲电信标准协会（ETSI）TC SmartM2M从几个应用领域的用例和要求着手，发展了M2M的通信架构和设备之间、网关之间、网络日志之间和提供M2M服务的应用之间的相关的接口，这项工作被引入了OneM2M。

ISO/IEC JTC 1/WG 7（传感器网络）已制定出ISO/IEC 29182服务标准，并用于传感网参考架构和协同信息处理的服务、接口。他们正研究通用应用和智能电网系统的传感器网络接口。

ITU设立了一个M2M中心组，研究物联网标准化工作并确定一般需求，其最初的重点放在医疗健康领域。JCA-IoT协调物联网的ITU-T工作，包括标识功能的网络部分和泛在传感网（USNs）。另外，ITU或多或少使得相关活动已发生了改变，如下一代网络，包括USN、安全和标识（命名和编号）。

除了802.15.4，IEEE也开展了智能传感器（1451系列）以及泛在的绿色社区控制（1888系列）的标准工作。信息模型，有时结合语义表述甚至本体论，已经被用在了不同的应用领域，如IEC TC 57的智能电网、IEC TC 65和ISO TC 184的工业自动化、ISO TC 205和ISO/IEC JTC 1/SC 25的楼宇自动化等。

物联网环境中的其他标准也很重要，就如IEC SC 3D所定义的产品数据标准，ISO、ITU所定义的标识标准，ISO/IEC JTC 1/SC 31及OGC所定义的定位标准。

Ⅳ. 物联网的应用

1. 无线传感网在智能交通的应用

智能交通中的无线传感在传统概念和无线传感网的设计需求等方面有所不同。多数情况下，传感器可以依赖一些基础设施供电，比如在这些系统中能效通常是次要的方面。无线传感网在智能交通的应用可细分为以下两类。

（1）固定的传感网，要么在车辆的主板上，要么作为交通基础设施的一部分。

（2）活动的传感网，单独的交通工具和其他移动的实体作为传感器。

后者包含一些应用，这些应用与物流、车辆和人的跟踪优化相关，而前者主要包含原来由有线传感器覆盖的应用。

智能交通管理解决方案依赖于精确的测量和对一个城市交通流量的可靠预测。这不仅包括对某街道的车辆密度或者某个公交车或火车上人数的估计，还包含对车辆和乘客起点及终点的分析。街道和十字路口的交通情形监控可以通过传统的有线传感器实现，如摄像机、感应环等。但是无线技术可以降低此类传感器的部署成本，它不会直接影响测量结果的准确度和有效性。

然而，通过扩展术语"传感器"的定义以及在很多车辆上和智能手机上利用无线技术，车辆本身和使用公共交通系统的乘客可以成为精确测量城市中交通流量的"传感器"。

城市物流是这个领域的另一个用例。城市化带来了很多挑战，尤其是在快速发展国家，因为大城市仍然在膨胀，而且日益富裕的人口导致进出城市中心的物流持续增长。

运输车辆的污染气体排放占城市空气污染的大部分，而且城市及其周围之间物流的流线型化是解决很多交通问题和改善空气质量的关键。如何降低由运输车辆造成的交通负荷，一个大有希望的方法就是引入城市整合中心（UCCs），即货仓处于城市之外，所有去往城市零售商的商品都先在这里整合，然后以最优路径运送，从而尽可能地充分利用卡车容量，也可减少车辆总数的需求量和运输商品的总距离。

要实现这样的优化，必须监控城市的交通流量和真实的商品流量以进行仔细的分析和规划。存在的挑战及解决方法和上面讨论的情形相似，但也有细微不同。车辆驶向城市时，至少必须实现pallet级水平的货物跟踪，而不是仅仅对车辆的子集进行跟踪。因此，托盘（或者其他包装单元）成为测量物流的"传感器"，另外多种无线技术（GPS, RFID, WLAN, cellular）与复杂数据分析技术的共同应用，从而能够获得优化配送调度和路由的数据，也在减少交通对环境的影响的同时，保障物流及时到达。

2. 物联网在智能电网的应用

电网不仅是电力行业的重要组成部分，也是国家持续发展的重要拼图。随着对电力依赖性的逐渐增长，世界范围内对电网可靠性和质量的要求也与日俱增。许多公用事业机构、研究机构和学者都在研究怎样实现电网的现代化，使之变得更加高效、清洁、安全、可靠，而且是交互式的。智能电网对新应用有着深远的影响：提供了在网络中安全集成更多可再生能源、电动车辆和分布式发动机的能力；通过需求响应、综合控制以及监控等功能，使得输电更加有效；利用自动电网重构防止或恢复电力中断（自愈能力）；支持用户对用电进行更多的控制而且积极地参与到电力市场。

传感器将成为智能电网发挥潜能的关键。"智能"电网背后的概念是能够响应实时需求。为了实现这一点，它就需要传感器提供实时信息。作为智能传感周边信息的无线传感网，它可以成为智能电网技术发展的重要途径。智能电网中的无线传感网技术也将进一步促使无线传感网在工业中的发展。

参考答案

1. 智能手机的应用程序可能是提供物联网世界的重要窗口，并且平板电脑和其他设备也将加入其中。

2. "物联网可以消除简单重复的任务，或创造以前不可能的东西，使更多的人完成更有价值的任务，让机器去做重复的工作"，格兰特·诺文说，他在伍德和道格拉斯公司负责市场营销，该公司生产物联网应用的无线通信硬件。

3. What is clear is that there is little point in two gadgets talking to each other if that doesn't bring an advantage to the owner, or some kind of efficiency gain.

4. That "double whammy" of population growth and ever-increasing urbanization mean that the IoT has the job of streamlining our cities, but in theory it will allow each of us to dispense with a lot of daily chores.

专业英语阅读策略与技巧

专业英语文献阅读是专业学习阶段的主要内容，学好科技英语会对从业者未来专业领域的发展和深造打下坚实的基础。但是科技英语在语法特点、词汇特点、句子和篇章文体特点等方面与普通英语都有很大的差异。因此，要学好科技英语，就必须了解这些特点，了解科技英语和普通英语的异同，从而在学习中克服障碍，运用正确的方法和策略。

1. 科技文献特点

科技英语（English for Science and Technology）和普通英语（General English）相比，在句子类型、句子长度、词汇、动词形式等方面都呈现出很大的差异。同时，科技英语文献的主要目的是传递信息、陈述事实、讲解过程、分析原理，而不是像文学作品那样表达自我、抒发感情。因此，科技英语呈现出以下一些突出的文体特点。

科技英语文献中通常包含3种词汇：一般词汇、半专业词汇和专业词汇。一般词汇即普通英语词汇或日常生活词汇，是科技英语中必不可少的词汇。半专业词汇由于其在不同学科或领域中有不同的意义，而且其复现率很高——在科技英语的文献中的出现率高达80%，因此，会给阅读带来一定的困难，是科技英语阅读中特别需要重视的词汇。专业词汇指那些在某一学科、某一领域或某一行业的专用术语或词汇。随着时代的发展和科技的不断进步，行业种类越来越多，学科门类越来越复杂，专业词汇也越来越多，新词不断涌现。科技英语中还有很多缩写词。如：sonar—sound navigation and ranging（声纳），comsat—communication satellite（通讯卫星），videophone—video telephone（可视电话）等。同时，随着科学技术的发展还将不断涌现出科技新词语。

大量使用第三人称是科技英语文体的一大特色。由于科技英语文献要体现客观公正、具有说服力，因此，应经常避免使用第一人称。另外，在学术性文章中，常常用"事"或"物"做句子的主语，用某种行为做主语或用it做主语，因此就会使用大量的被动语态的句子。

简单明了也是科技文献的特色。为了使信息传递准确、快捷，作者往往把要表达的内容系统地、直截了当地、简单明快地表达出来，最大限度地增加可读性，减少读者理解方面的困难。

2. 科技英语阅读的技巧和方法

整体理解文章，抓段落大意。在阅读时，同样可以采用一般英语阅读中所使用的略读（skimming）、寻读（scanning）、研读（study reading）等阅读技巧。科技英语文献资料的写作一般都比较规范，在阅读时只要认准了主题句，就抓住了该段的段意，从而在阅读时避免细研语法、琢磨词汇，既可注重整体理解，同时又可以提高阅读速度，节省时间。

克服词汇障碍。科技英语词汇和普通英语词汇有很大的不同，难度较大，不好理解。扩大词汇量是提高阅读能力的一条切实可行的重要途径。同时也要提高认识常用的词根和词缀的能力，就能辨认大量不熟悉的同族词。

从专业基础下工夫，克服背景知识方面的障碍。通过文献检索，相关论文阅读等方式，扩展背景知识去理解阅读材料所传递的信息，也就是用不断充实的新知识去理解、消化、吸收材料的内容。

Unit 2

Architectures of IoT

Passage A IoT Reference Architectures

Passage B Sensor Network Reference Architecture

Passage C Industrial Internet Reference Architecture

Passage A IoT Reference Architectures

I. Introduction

A number of standards organizations have worked in the area of Internet of Things. There are some application domain specific architectures, and some more generic reference architectures. Additionally a number of fora and consortia have been active in proposing architectures for the Internet of Things, some international research projects have also worked on such developments. Since there is no universally accepted definition of IoT, different groups have developed different approaches according to the domain in which they are active and where IoT technology (or aspects thereof) is appropriate. Currently there is IoT related activity in JTC 1 SCs (System committees) and other entities such as WG7 (Working Group 7), SC6, SC25, SC27, SC29, and SC31 with cloud aspects also in SC38. In ISO other TCs (Technology committees) involved in aspects of IoT include TC104, TC122, TC211, TC215, etc. In IEC other TCs and systems committees involved in IoT include IEC TC100 and the systems committees on Smart Cities. In ITU-T there is a Joint Coordination Activity on IoT and various questions related to IoT are being progressed in SG9 (Study Group 9), SG11, SG13, SG16, SG17, and in ITU-R there is work in SG5. In ETSI there is significant activity on M2M. Fora and consortia involved in IoT related standardization (with an example of their IoT related specifications) include OMA (Web Services Network Identity), 3GPP (M2M) ECMA (e.g. ECMA-262), OGC (Sensor Web), IEEE (802.24 TAG) and many others. A single IoT reference architecture suitable for all these bodies is not an achievable goal so a number of IoT reference architectures are defined.

II. Requirements for IoT Reference Architecture

The Reference Architecture should not prescribe any specific method for implementation unless this is a requirement for a specific application domain. The architecture should describe the system or systems for the IoT, the major components involved, the relationships between them, and their externally visible properties.

The performance level of each requirement may be important, and it may involve not only the performance of a specific component but the overall system performance. In some applications specific requirements are critical for successful implementation.

General requirements for IoT Reference Architecture are as follows:

- Regulation Compliance
- Autonomous Functionality
- Auto-configuration
- Scalability
- Discover Ability
- Heterogeneity
- Unique Identification
- Usability
- Standardized Interfaces
- Well Defined Components

- Network Connectivity
- Timeliness
- Time-Awareness
- Location-Awareness
- Context-Awareness
- Content-Awareness
- Modularity
- Reliability
- Security
- Confidentiality and Privacy
- Manageability
- Risk Management

Application categories may overlap several domains or may be domain specific. Examples of requirements only needed or critical in specific applications are:

(1) Power and Energy Management

The IoT RA should support power and energy management particularly in networks with battery powered components. Different strategies will suit different applications but may include using low-power components in devices, limiting the communication range, limiting the local processing and storage capacity, supporting sleep modes and energy harvesting.

(2) Accessibility

The IoT RA should support accessibility preferences and requirements. In some application domains accessibility will be important for the usability of the IoT system and wherever there is significant user involvement in the configuration, operation and management of the system.

(3) Human Body Connectivity

The IoT RA should support implementations involving safe human body connectivity. In order to provide communication capabilities related with the human body in compliance with regulation and laws a special quality of service is required, reliability and security have to be quantified, and privacy protection is required.

(4) Service Related Requirements

The IoT RA should support service related requirements such as prioritisation, semantic services, service composition, usage tracking, and service subscription which will vary according to the application domain and implementation. For example Location-Awareness in some applications may need the support of location-based services with specified accuracy.

(5) "Environmental Impact" Requirements

The IoT RA should support components, services and capabilities which lead to minimal "environmental impact" for an implementation.

Example of a specific set of requirements for a specific application domain: those identified for sensor web applications.

- Sensors will be Web accessible;

- Sensors and sensor data will be discoverable;
- Sensors will be self-describing to humans and software (using a standard encoding);
- Most sensor observations will be easily accessible in real time over the Web;
- Standardized Web services will exist for accessing sensor information and sensor observations;
- Sensor systems will be capable of real-time mining of observations to find phenomena of immediate interest;
- Sensor systems will be capable of issuing alerts based on observations, as well as be able to respond to alerts issued by other sensors;
- Software will be capable of on-demand geolocation and processing of observations from a newly-discovered sensor without a priori knowledge of that sensor system;
- Sensors, simulations, and models will be capable of being configured and tasked through standard, common Web interfaces;
- Sensors and sensor networks will be able to act on their own (i.e. be autonomous).

III. ISO/IEC 30141 Project

ISO /IEC JTC1 WG10 (IoT) - the international standard work group for IoT, is drafting the IoT Reference Architectures international standard, which is numbered ISO/IEC 30141. This part will introduce some contents of the standard. This ISO/IEC 30141 standard specifies the general IoT Reference Architecture in terms of defining Characteristics, the IoT Conceptual Model, Reference Model, and Architecture Views and Reference Architecture from different architectural views, and the common entities, and high-level interfaces connecting the entities.

IoT is defined as an infrastructure of interconnected objects, people, systems and information resources together with the intelligent services to allow them to process information of the physical and the virtual world and react. The IoT Reference Architecture (IoT RA) represented in this International Standard provides a Conceptual model and Reference model. The IoT RA provides a generic description as a basis when to develop an IoT system architecture that will have specific system requirements that should be met and the specific system requirements can vary from one IoT system to another.

The IoT RA serves the following goals, to describe:
— the Conceptual Model (CM), including the core characteristics of an IoT system;
— the Reference Model (RM), including the views and domains of an IoT system.

The IoT RA supports the following important standardization objectives:
— to provide a technology-neutral reference point for defining standards for IoT;
— to encourage openness and transparency in the development of an IoT system architecture.

The conceptual model (CM) provides a common structure and definitions for describing the concepts of, and relationships among, the entities within IoT systems. It must be generic, abstract and simple. In order to achieve this goal, it is important to clarify the fundamentals of the IoT systems by asking the following questions:
- ➢ What is the big picture of the overall IoT entities and their relationships?
- ➢ What are the key concepts in a typical IoT system?
- ➢ What are the relationships between the entities, especially between digital entities and their physical

entities?
- Who and where are the actors?
- How the things and services collaborate via the network?

The model diagram in Figure 2.1 provides the big picture of all key IoT entities defined in this CM, their relationships and their interactions. The IoT-User can be human (human user) or non-human (digital user) such as robots or automation services, which act on behalf of human users. Digital user consumes remote services through endpoint, which is attached to the communication network, while a human user interacts through applications, which are implemented by a set of components. Some of the components of application can communicate with remote services via network using exposed endpoint. Physical entity here is the thing which is to be controlled by an actuator or monitored by a sensor. A virtual entity represents a physical entity in the IT world. Both actuators and sensors are kinds of IoT devices. IoT devices may interact through an endpoint to have network communicate directly or connect with an IoT gateway first if itself does not have communication capabilities. Services are implemented by components and they can be located anywhere. Services can be discovered and consumed via different types of networks through endpoints.

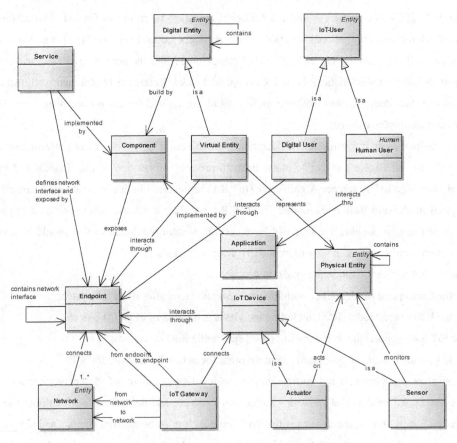

Figure 2.1　Big picture of all key IoT entities defined in this CM.

A reference model (RM) is an abstract framework for understanding significant relationships among the entities of an environment, and for the development of consistent standards or specifications supporting that environment. A reference model is based on a small number of unifying concepts and may be used as a basis for education and explaining standards to a non-specialist. A reference model is not directly tied to any standards, technologies or other concrete implementation details, but it does seek to provide common semantics that can be used unambiguously across and between different implementations.

There are a number of concepts rolled up into that of a reference model. The RM is abstract, and it provides information about environments of a certain kind. A RM describes the type or kind of entities that may occur in such an environment, not the particular entities that actually do occur in a specific environment. A RM describes both types of entities or domains (things that exist) and their relationships (how they connect, interact with one another and exhibit joint properties). A list of entity types, by itself, doesn't provide enough information to serve as a reference model. A RM does not attempt to describe "all things." A RM is used to clarify "things within an environment" or a problem space. To be useful, a RM should include a clear description of the problem that it solves, and the concerns of the stakeholders who need the problem to be solved. A RM is technology agnostic. A RM's usefulness is limited if it makes assumptions about the technology or platforms in place in a particular computing environment. A RM typically is intended to better understand a class of problems, not to provide specific solutions for those problems. As such, it must aid the process of inventing and evaluating a variety of potential solutions in order to assist the practitioner. The RM is useful:

— to create standards for both the objects that inhabit the model and their relationships to one another;
— to educate;
— to improve communication between people;
— to create clear roles and responsibilities;
— to allow the comparison of different entities.

The reference architecture can be understood as contexts provided with common features, vocabulary, requirements together with supporting artifacts to enable their use.The artifacts are the description of the major foundational architecture components that provide guidelines and constrains for instantiating solution architectures.The solution architectures can be defined not only from a different emphasis or viewpoint but also at many different levels of detail and abstraction; they consist of a list of entities and functions and some indication of their connections, interrelations and interactions with each other and with functions located outside of predefined architecture patterns representing the entities and functions.

Figure 2.2 shows the architecture continuum from the CM through the entity-based RM and domain-based RM to a number of different views of the RA. The consistent architecture continuum should be maintained not only in this hierarchy (e.g., CM、RM、RA) but also in evolutionary updates over time and it should be clearly understood through clear documentation of the architecture descriptions.

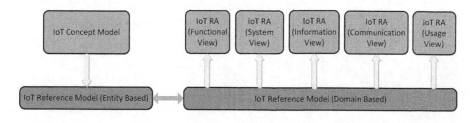

Figure 2.2　Relation between IoT concept model, reference model, and reference architectures.

In this standard, after examining various kinds of deployed IoT systems and developing the IoT Conceptual Model (CM) through IoT system decomposition, common and representative domains of IoT systems are identified by focusing on the IoT systems' stakeholders and hardware/software. Using these common and representative domains provides an effective and representative Reference Model (RM) of the IoT systems for the various purposes and uses of the RM. The ISO/IEC 30141 Project is expected to be finished by the mid of 2017.

Ⅳ. New Words

fora ['fɔ:rə]	n.	论坛；讨论会
consortia [kən'sɒti:əm]	n.	社团；银行团
coordination [kəʊ'ɔ:dɪ'neɪʃn]	n.	协调；和谐
conceptual [kən'septʃuəl]	adj.	观念的；概念的
heterogeneity [ˌhetərə'dʒə'ni:ətɪ]	n.	多向性；异质性
semantic [sɪ'mæntɪk]	adj.	语义的；语义学的
geolocation [dʒɪɒləʊ'keɪʃn]	n.	地理定位

Ⅴ. Phrases

auto-configuration	自动配置；自动设定
self-describing	自描述；自我描述

Ⅵ. Abbreviations

TC	Technology committees	技术委员会
ITU-T	International Telecommunication Union-Telecommunication Sector	国际电信联盟—电信标准化局
ITU-R	International Telecommunications Union-Radio Communications Sector	国际电信联盟—无线通信委员会
M2M	Machine to Machine	机对机通信
OMA	Open Mobile Architecture	开放式移动体系结构
3GPP	the 3rd Generation Partnership Project	第三代合作伙伴项目
OGC	Open GIS Consortium	开放地理空间信息联盟
RM	Reference Model	参考模型
CM	Conceptual model	概念模型

Passage B Sensor Network Reference Architecture

The ISO/IEC 29182 series focuses on a generic architecture for sensor networks, which has been developed by JTC 1 WG7 – Sensor Networks. The purpose of the standard is to (i) provide guidance to facilitate the design and development of sensor networks,(ii) improve interoperability of sensor networks, and (iii) make sensor network components plug-and-play, so that it becomes fairly easy to add/remove sensor nodes to/from an existing sensor network. It can be used by sensor network designers, software developers, system integrators, and service providers to meet customer requirements, including any applicable interoperability requirements. The seven parts of the 29182 series are described below.

Ⅰ. ISO/IEC 29182-1:2013 Information technology — Sensor networks: Sensor Network Reference Architecture (SNRA) — Part 1: General overview and requirements

Part 1 provides a general overview of the characteristics of a sensor network and the organization of the entities that comprise such a network. It also describes the general requirements that are identified for sensor networks. It presents sensor network architecture from the communications and service provisioning perspectives (standalone, interconnected, connected to other networks), the characteristics of sensor networks that differentiate them from traditional data networks, and general requirements for sensor networks.

Ⅱ. ISO/IEC 29182-2:2013 Information technology — Sensor networks: Sensor Network Reference Architecture (SNRA) — Part 2: Vocabulary and terminology

Part 2 provides a general overview of the characteristics of a sensor network and the organization of the entities that comprise such a network. It also describes the general requirements that are identified for sensor networks. It presents terms and definitions for selected concepts relevant to the field of sensor networks, establishes a general description of concepts in this field, and identifies the relationships among those concepts.

Ⅲ. ISO/IEC 29182-3: Information technology — Sensor networks: Sensor Network Reference Architecture (SNRA) — Part 3: Reference architecture views

ISO/IEC 29182-3:2014 provides Sensor Network Reference Architecture (SNRA) views. The architecture views include business, operational systems, and technical perspectives, and these views are presented in functional, logical, and/or physical views where applicable. ISO/IEC 29182-3:2014 focuses on high-level architecture views which can be further developed by system developers and implementers for specific applications and services. Sensor network functional architecture is shown as Figure 2.3.

Ⅳ. ISO/IEC 29182-4: 2013 Information technology — Sensor networks: Sensor Network Reference Architecture (SNRA) — Part 4: Entity models

Part 4 presents models for the entities that enable sensor network applications and services according to the Sensor Network Reference Architecture (SNRA). It provides basic information about and high-level models for various entities that comprise a sensor network. Physical entities are pieces of hardware and actual devices or components thereof that form the network, such as sensor nodes and gateways while functional entities represent certain tasks that may be carried out on one or more types of physical entity.

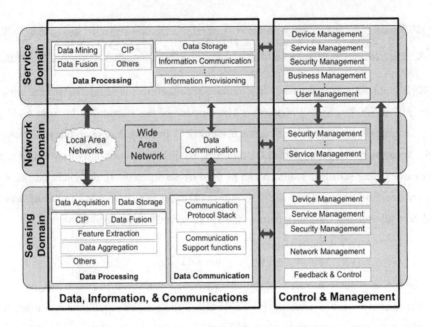

Figure 2.3 Sensor network functional architecture

V. ISO/IEC 29182-5:2013 Information technology — Sensor networks: Sensor Network Reference Architecture (SNRA) — Part 5: Interface Definitions

Part 5 provides the definitions and requirements of sensor network (SN) interfaces of the entities in the Sensor Network Reference Architecture and covers the following aspects:

- interfaces between functional layers to provide service access for the modules in the upper layer to exchange messages with modules in the lower layer;
- interfaces between entities introduced in the Sensor Network Reference Architecture enabling sensor network services and applications.

Figure 2.4 depicts all the interfaces addressed in this part.

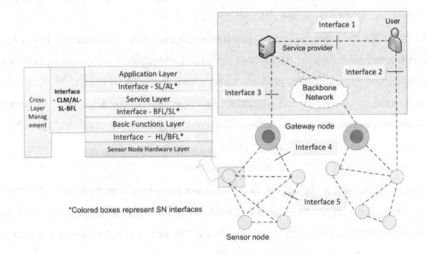

Figure 2.4 Various interfaces of a sensor network

Ⅵ. **ISO/IEC 29182-6: Information technology — Sensor networks: Sensor Network Reference Architecture (SNRA) — Part 6: Applications**

This part provides a compilation of sensor network applications for which International Standardized Profiles (ISPs) are needed, guidelines for a structured description of sensor network applications, examples of structured sensor network application descriptions such as management of mobile assets in hospitals and container monitoring in the supply chain, that support the development of ISPs in a follow up step.

Ⅶ. **ISO/IEC 29182-7: Information technology — Sensor networks: Sensor Network Reference Architecture (SNRA) — Part 7:Interoperability Guidelines**

This Part provides a general overview and guidelines for achieving interoperability between service providers and related entities in a heterogeneous sensor network as illustrated below.

Ⅷ. **New Words**

integrators ['ɪntɪgreɪtə]   n. 集成商
terminology [ˌtɜːmɪˈnɒlədʒi]   n. 专门名词，术语，术语学；用词
gateway ['geɪtweɪ]   n. 网关
interface ['ɪntəfeɪs]   n. 接口；交界面；界面
interoperability ['ɪntərɒpərə'bɪlətɪ] n. 互用性；协同工作的能力
heterogeneous [ˌhetərə'dʒiːniəs] adj. 异构的

Ⅸ. **Abbreviations**

SNRA   Sensor Network Reference Architecture   传感器网络参考架构
ISPs   International Standardized Profiles   国际标准化轮廓

Passage C Industrial Internet Reference Architecture

The Industrial Internet Consortium (IIC) was founded in March 2014 to bring together the organizations and technologies necessary to accelerate the growth of the Industrial Internet by identifying, assembling and promoting best practices. Membership includes small and large technology innovators, vertical market leaders, researchers, universities and government organizations.

The goals of IIC include:
- Driving innovation through the creation of new industry use cases and testbeds for real-world applications;
- Defining and developing the reference architecture and frameworks necessary for interoperability;
- Influencing the global development standards process for internet and industrial systems;
- Facilitating open forums to share and exchange real-world ideas, practices, lessons, and insights;
- Building confidence around new and innovative approaches to security.

An Industrial Internet Reference Architecture Technical Report has been released by IIC AT 2015. The contents of the report will be discussed in this part.

The Industrial Internet is an internet of things, machines, computers and people, enabling intelligent industrial operations using advanced data analytics for transformational business outcomes. It embodies the convergence of the global industrial ecosystem, advanced computing and manufacturing, pervasive sensing and ubiquitous network connectivity.

There are many interconnected systems deployed today that combine hardware, software and networking capabilities to sense and control the physical world. These industrial control systems contain embedded sensors, processors and actuators that provide the capability to serve specific operational or business purposes, but, by and large, these systems have not been connected to broader systems or the people who work with them.

The Industrial Internet concept is one that has evolved over the last decade to encompass a globally interconnected network of trillions of ubiquitous addressable devices and collectively representing the physical world.

Ⅰ. Industrial Internet Reference Architecture

An industrial internet reference architecture provides guidance for the development of system, solution and application architectures. It provides common and consistent definitions in the system of interest, its decompositions and design patterns, and provides a common vocabulary with which to discuss the specification of implementations so that options may be compared.

The Industrial Internet Reference Architecture (IIRA) is a standard — based open architecture for IISs. To maximize its value, the IIRA has broad industry applicability to drive interoperability, to map applicable technologies, and to guide technology and standard development. The description and representation of the architecture are generic and at a high level of abstraction support the requisite broad industry applicability. The IIRA distills and abstracts common characteristics, features and patterns from use cases well understood at this time, prominently those that have been defined in the Industrial Internet Consortium (IIC). The IIRA design transcends today's available technologies and in so doing is capable of identifying technology gaps based on the architectural requirements. This will in turn drive new technology development efforts by the Industrial Internet community.

When considering complex systems such as those expected of Industrial Internet Systems (IISs), many stakeholders are involved. These stakeholders have many intertwining concerns pertinent to the system of interest. Their concerns cover the full lifecycle of the system, and their complexity calls for a framework to identify and classify the concerns into appropriate categories so that they can be evaluated and addressed systematically.

To address this need, the Industrial Internet Consortium has defined an architecture framework that describes the conventions, principles and practices for the description of architectures established within a specific domain of application and/or community of stakeholders.

II. Industrial Internet Viewpoints

The various concerns of an IIS are classified and grouped together as four viewpoints as shown in Figure 2.5:

- Business
- Usage
- Functional
- Implementation

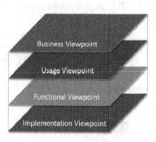

Figure 2.5 Architecture Viewpoints

The Business Viewpoint attends to the concerns of the identification of stakeholders and their business vision, values and objectives in establishing an IIS in its business and regulatory context. It further identifies how systems built on the IIS achieve the stated objectives through its mapping to fundamental system capabilities. These concerns are business-oriented and are of particular interest to business decision-makers, product managers and system engineers.

Business-oriented concerns such as value proposition, expected return on investment, cost of maintenance and product liability must be evaluated when considering an Industrial Internet System as a solution to business problems. To identify, evaluate and address these business concerns, IIC introduces a number of concepts and define the relationships between them, as shown in Figure 2.6.

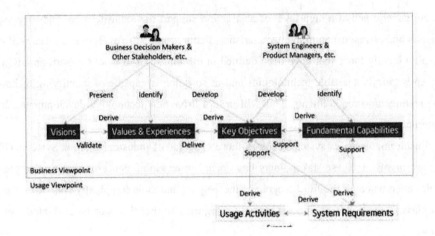

Figure 2.6 Business Viewpoint

The usage viewpoint addresses the concerns of expected system usage. It is typically represented as sequences of activities involving human or logical users that deliver its intended functionality in ultimately achieving its fundamental system capabilities.The stakeholders of these concerns typically include system engineers, product managers and the other stakeholders including the individuals who are involved in the specification of the IIS under development and who represent the users in its ultimate usage.

The usage viewpoint is concerned with how an Industrial Internet System realizes the key capabilities identified in the business viewpoint. The usage viewpoint describes the activities that coordinate various units of work over various system components. These activities-describing how the system is used-serve as an input for system requirements including key system characteristics and guide the design, implementation, deployment, operations and evolution of the IIS as shown in Figure 2.7.

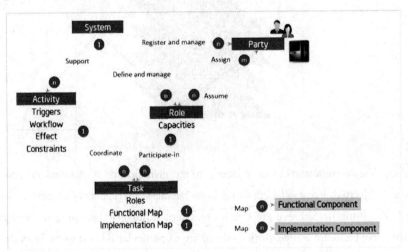

Figure 2.7 Role, Party, System, Activity and Task

The functional viewpoint focuses on the functional components in an IIS, their interrelation and structure, the interfaces and interactions between them, and the relation and interactions of the system with external elements in the environment, to support the usages and activities of the overall system.These concerns are of particular interest to system and component architects, developers and integrators.

IIC creates the concept of functional domain to address the key concerns surrounding the functional architecture of Industrial Internet Systems. A Functional Domain is a top-level functional decomposition of an Industrial Internet System, each providing a predominantly distinct functionality in the overall system.

A use case under consideration and its specific system requirements will strongly influence how the functional domains are decomposed, so in a concrete architecture derived and extended from this reference architecture, additional functions may be added, some of the functions described here may be left out or combined and all may be further decomposed as needed.

IIC decompose a typical IIS into five functional domains as shown in Figure 2.8:
- Control domain
- Operations domain
- Information domain
- Application domain
- Business domain

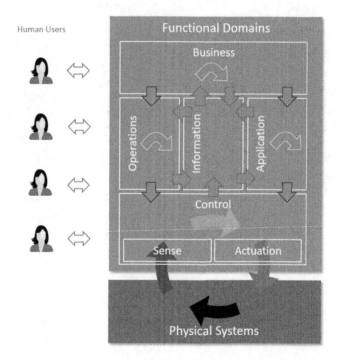

Figure 2.8 Functional domains

The implementation viewpoint deals with the technologies needed to implement functional components, their communication schemes and their lifecycle procedures. These components are coordinated by activities (Usage viewpoint) and supportive of the system capabilities (Business viewpoint). These concerns are of particular interest to system and component architects, developers and integrators, and system operators.

The implementation viewpoint is concerned with the technical representation of an Industrial Internet System and the technologies and system components required to implement the activities and functions prescribed by the usage and functional viewpoints.

Ⅲ. New Words

testbeds [testbed]   n. 测试平台
interoperability [ˌɪntərɒpərə'bɪlətɪ]   n. [计]互操作性；互用性
addressable [ə'dresəbl]   adj. 可寻址的
distill [dis'til]   vt. 提取
specification [ˌspesɪfɪ'keɪʃ(ə)n]   n. 规格；说明书

Ⅳ. Phrases

business-oriented   面向业务

Ⅴ. Abbreviations

IIC   Industrial Internet Consortium   工业互联网联盟
IIRA   Industrial Internet Reference Architecture   工业网络参考架构
IIS   Industrial Internet Systems   工业网络系统

Exercises

Translate the following sentences into Chinese or English.

1. Radio frequency identification (RFID) technology has been widely used in the field of construction during the last two decades. Basically, RFID facilitates the control on a wide variety of processes in different stages of the lifecycle of a building, from its conception to its inhabitance.

2. Based on the research of recent IoT reference architecture, we differentiate and analyze some network conceptions closely related to the IoT respectively, such as Ubiquitous Networks, Sensor Networks and M2M, which contrasts some architectures of the IoT based on the existing ones.

3. 它的目的是，使终端用户想要使用物联网技术，使管理者有兴趣了解由这些新技术激发的机会，使系统架构师对基本模型的概述感兴趣。

4. 2015物联网/ CPS安全研讨会的目标是，汇集国际领先的学术和工业，努力鉴定和讨论主要的技术挑战及最新研究结果，解决物联网与CPS安全和隐私方面的研究。

参考译文 Passage A 物联网参考架构

Ⅰ. 简介

大量标准组织已致力于物联网领域研究，提出了一些应用于特定领域的体系架构，及更通用的参考架构。另外，一些论坛和联盟也一直在积极地为物联网提出架构，一些国际研究项目也取得了进展。由于没有统一的物联网定义，不同团体根据他们从事的不同领域以及物联网技术适用的领域，提出了不同的观点。目前在系统委员会（JTC 1 SCs）中物联网相关的工作组包括WG7、SC6、SC25、SC27、SC31和云方向（SC38）。在国际标准组织（ISO）中，涉及到物联网方面的其他技术委员会（TC）包括TC104、TC122、TC211及TC215等。在IEC中，物联网涵盖的TC和SC，包括IEC TC100及智能城市系统委员会。国际电联（ITU-T）中基于物联网的联合协调活动和物联网相关的各种问题正在SG9、SG11、SG13、SG16、SG17中推进，国际电联无线电部门（ITU-R）SG5中也在研究这一课题。欧洲电信标准协会（ETSI）有关于M2M的重要活动。关于物联网标准化的论坛和联盟（以物联网相关规范为例）包括Web服务网络识别（OMA）、3GPP（M2M）CMA（如ECMA-262）、传感器网络（OGC）、802.24技术咨询组（IEEE）等。要使单一的物联网参考架构适用于这些多有的机构，是个不能实现的目标，所以出现了一批物联网参考架构。

Ⅱ. 对物联网参考架构的需求

参考架构不应该规定任何具体的实施方法，除非这是一个特定应用领域的要求。该体系架构应描述该物联网体系，它们所涉及的主要组成部分，彼此的关系以及外部的可见属性。

需求的性能水平可能是重要的，它可能不仅涉及一个特定组件的性能，而且还涉及系统整体的性能。在某些应用程序中，特定的需求是成功实施应用的关键。对于物联网参考架构的通用要求包括：

- 法令遵循
- 自治功能
- 自动配置
- 可扩展性
- 可发现性
- 异构性
- 唯一标识
- 可用性
- 标准化接口
- 定义良好的组件
- 网络连通性
- 时效性
- 时间感知
- 位置感知
- 环境感知
- 内容感知

- 模块性
- 可靠性
- 安全性
- 保密性
- 易管理性
- 风险管理

应用程序类别可以覆盖多个域或某个特定域。在特定应用领域关键需求的例子如下。

（1）电力及能源管理

物联网参考架构支持电力及能源管理，尤其是在由电池供电的网络。不同策略将适用于不同的应用，但可能包括使用低功率组件的设备，有限的通信范围，有限的本地处理和存储容量，支持睡眠模式及能量收集。

（2）可访问性

物联网参考架构支持可访问性需求和要求。在某些应用领域，可访问性对物联网系统的可用性是重要的，而且对于系统配置、运行和管理中有重要用户涉及的地方也是很重要的。

（3）人体连通性

物联网参考架构应支持实现关于安全的人体连接的操作。为了提供与人体相关的通信功能，特殊服务质量必须符合法律法规，可靠性和安全性必须进行量化，并要求保护隐私。

（4）相关服务要求

物联网参考架构应该支持与服务相关的要求，如优先性、语义服务、服务组合、使用跟踪、订阅服务，这些要求将根据不同的应用领域和实现而变化。例如，某些应用的位置感知需要基于特定精度的位置服务来支持。

（5）"环境影响"的要求

物联网参考架构应支持组件、服务和在实现应用时候产生最小"环境影响"的能力。特定应用领域中，一组特定要求的示例（用于确定传感器网络的应用程序需求）包括：

- 传感器网络的Web可访问性；
- 传感器和传感器数据的可发现性；
- 传感器向人类和软件（使用标准编码）的自描述；
- 大多数传感器能在实时的Web中被访问；
- 使用标准的Web服务访问传感器信息和进行传感器观测；
- 传感器系统有能力对观察值进行实时挖掘从而寻求切身利益；
- 传感器系统能基于观测发出警报，也能响应其他传感器发出的警报；
- 软件能按需定位和处理观测，没有传感器系统的先验知识的条件下，软件可以对来自新发现的传感器观察值进行按需定位和处理；
- 传感器、仿真和模型可配置而且可通过标准、通用Web接口分配任务；
- 传感器和传感器网络能响应自己的行为（即自治）。

Ⅲ. ISO/IEC 30141项目

ISO/IEC JTC1 WG10 (IoT)（国际标准化组织/国际电工委员会的第一联合技术委员会）是一个国际物联网标准工作组，起草物联网国际标准参考架构，其编号为ISO/IEC 30141。本部分将介绍一些关于该标准的内容。ISO/IEC 30141标准规定了通用物联网参考架构，依据特性定义了参考架构，来自不同架构观念的物联网概念模型、参考模型、架构观及参考架构，通用实体和连接实体的高层次界面。

物联网被定义为一种互联对象、互联人体、互联系统和互联信息资源所组成的基础设施，可提供处理真实世界和虚拟世界信息，并作出回应的智能服务。此项国际标准所表述的物联网参考架构（IoT RA）提供了一个概念模型和参考模型。当联网系统架构发展，可以满足从一个物联网架构改变到另一个物联网架构的具体要求时，物联网参考架构将提供一种通用的描述作为基础。

物联网参考架构服务于以下目标：
— 概念模型（CM）包括物联网系统的核心特点；
— 参考模型（RM）包括物联网系统的视图和域；

物联网参考架构支持以下重要的标准化目的：
— 为物联网定义标准提供与技术中立的参考点；
— 支持物联网系统架构发展公开透明化。

概念模型（CM）提供共同的结构和定义，用于描述概念及描述物联网系统中实体之间的关系。
➤ 整个物联网实体及其相互关系的整体情况是什么？
➤ 在典型的物联网系统中，其关键概念是什么？
➤ 实体之间的关系，尤其是数字实体和物理实体之间的关系是什么？
➤ 这些角色在哪儿？是什么？
➤ 事情与服务如何通过网络合作？

图2.1中的模型图提供所有关键的物联网实体、彼此关系及相互作用实体的整体情况，其中实体在概念模型（CM）中定义。物联网用户可以是人类（人用户）或非人类（数字用户），如机器人或自动化服务，这代表人用户。数字用户通过终端来消耗远程服务，并连接到通信网络。同时，人类用户通过由一组组件实施的应用程序进行交互作用。应用程序的一些组件可以通过网络使用暴露的端点进行远程服务通信。这里的物理实体是由一个执行器控制或传感器监测的。在信息科技（IT）世界，一个虚拟的实体代表一个物理实体。执行器和传感器都是物联网设备。物联网设备可以通过一个端点进行交互作用，以直接地进行网络通信，或在其自身没有通信能力的前提下首先连接到物联网关。由组件实施的服务可位于任何地方。通过端点的不同类型的网络可发现和消耗服务。

参考模型（RM）是一个抽象的框架，此框架被用于了解一个环境中的实体之间的重要关系，也用于发展为支持此环境而制定的标准或规范。参考模型是基于小数量的统一概念，可以用作教育的基础和非专业人员的解释标准。参考模型并没有直接关系到不同实现之间的任何标准、技术或其他具体的实施细节，但它试图寻求共同的语义，以明确地运用于不同的应用层。

一系列概念累积起来成为一个参考模型。参考模型是抽象的，提供了一种特定环境的信息。参考模型描述的是可能在这样的环境下发生的一类实体，而不是确确实实发生在特定的环境下的特殊实体。参考模型不仅描述实体或域（存在的事物）的类型，还描述它们之间的关系（如何连接，如何交互作用，以及如何表现出共同的属性）。实体类型列表本身并没有提供足够的信息作为参考模型。参考模型并不能描述"一切事物"，却可用它来阐明"一定条件下的事物"或问题空间。为了更方便使用，参考模型应包括所要解决问题的清晰描述以及利益相关者的关注点，利益相关者需用参考模型解决问题。参考模型对物联网涉及的技术是无法预知的。如果假设在一个特定的技术或平台计算环境下，参考模型的用途就会有限。模型通常旨在促进理解问题，而不会针对这些问题提供具体的解决方案。因此，它必须在发明和评估各种可能的解决方案的过程中协助操作人员。模型的作用如下：
— 为建立模型的对象及他们之间的关系创建标准；
— 教育；
— 提高人与人之间的沟通；
— 创建明确的角色和职责；
— 允许不同实体的比较。

参考架构可以理解为这样的内容，这些内容有共同特征、词汇和要求以及可供使用的工件。支持工件主要是对基础架构组件的一些描述，为实例化方案架构提供指导和约束。该解决方案不仅可以从不同的侧重点和观点来定义，也可以从许多不同层次的细节和抽象上来定义；它们由一个实体和功能的列表，一些相互连接、相互关系、相互作用的指示，以及与代表实体和功能的境外预设功能架构模式组成。

从图2.2可看出，从概念模型到基于实体和基于域的参考模型，再到对参考模型的一系列不同观点。一致的体系结构连用体不仅应保持在这个层次（如概念模型、参考模型、参考架构），也应随着时间的推移进化更新，并通过对结构描述清晰的文档得到清楚的理解。

在这一标准中，通过分解物联网系统，研究各种部署的物联网系统和发展物联网的概念模型（CM），物联网系统共同的、有代表性的领域将重点关注物联网系统的利益相关者和硬件/软件。这些共同的、有代表性的领域，可为各种用途的参考模型提供有效的、有代表性的物联网系统参考模型（RM）。

ISO/IEC 30141项目预计将于2017年中期完成。

参考答案

1. 射频识别在过去的20年已被广泛地应用于建筑领域。它几乎对建筑周期的每个阶段都有促进作用，从起初的概念到最后的实实在在的住房。

2. 基于已有的物联网参考体系架构，我们分别辨析了泛在网、传感网、M2M等几个与互联网密切相关的网络概念。

3. It is aimed at end users who want to use IoT technologies, managers interested in understanding the opportunities generated by these novel technologies, and system architects who are interested in an overview of the underlying basic models.

4. The goal of the 2015 IoT/CPS-Security workshop is to bring together internationally leading academic and industrial researchers in an effort to identify and discuss the major technical challenges and recent results aimed at addressing all aspects of security and privacy for IoT and CPS.

科技翻译的标准

无论是在东方还是在西方，传统的翻译理论历来强调一个"信"字，也就是说，将"忠实"视为翻译的基本标准。目前，科技翻译还没有一个可资遵循的统一标准。因此，我们根据翻译理论和科技语言的特点，在总结前人提出的翻译标准的基础上，给出以下标准，供读者参考。

1. 正确无误

众所周知，科学技术对于准确性的要求特别严格，大到理论的阐述，小到数据的举证，都不能有丝毫的谬误和误差。这就要求科技翻译也必须做到准确无误。鉴于此，"准确"是科技翻译的首要标准。它着眼于译文的内容，不能有任何模棱两可之处。例如：

There are always crossed transverse steady and longitudinal alternating fields.

译文1：始终存在着交叉横向稳定和纵向交变磁场。

译文2：始终存在着交叉的横向稳定磁场和纵向交变磁场。

分析：译文1的意思不够明确，容易让人把"交叉""横向"和"稳定"视为并列关系，而且也与"纵向交变"并列，共同修饰"磁场"一词，让人误以为磁场具有两种特征。译文2增加了一个"的"字，清楚地表明"交叉"修饰前后两个磁场，又重复翻译了"磁场"这个词，和原文名词fields对等，使语义更明确。

2. 通达顺畅

这是指译文应该符合目标语的语言规则和行文习惯，读起来通顺畅达，可读易懂。"顺达"是保证"准确"的基本条件，它要求词语的选择、组合和搭配要恰到好处。句子的语序要恰当排列，各句之间语义逻辑紧密衔接，句型结构流畅，能准确地传达原语的情态、时态、语态等，并能恰当体现语义的重点。例如：

This possibility was supported to a limited extent in the tests.

译文1：在实验中这一可能性在有限的程度上被支持了。

译文2：这一可能性在实验中于一定程度上得到了证实。

分析：译文1似乎达到了"准确"的要求。但是，翻译为被字句不符合汉语的表达习惯。故其既不准确，又不通顺，即没有做到"顺达"。译文2真正做到了"准确"和"顺达"。

3. 简单干练

科技英语从本质上说就是实用英语，所以译文要尽可能简练，即简短而凝练，没有冗词废字。换句话说，应在准确、顺达的基础上，力求简洁明快、精炼概要。例如：

Each product must be produced to rigid quality standard.

译文：每件产品均须达到严格的质量标准。

分析：如果把这个句子译为"每件产品都必须生产得符合严格的质量标准"就显得生硬、啰嗦。

4. 规范统一

这是指，语言、文字、术语、简称、符号、公式、语体和计量单位都要规范统一，符合国家和国际标准。

例如：generation gap代沟（试比较：世代隔阂），full-time scientists专职科学家（试比较：全部时间科学家），part-time scientists兼职科学家（试比较：部分时间科学家）。

Unit 3

RFID

Passage A   An Overview of RFID

Passage B   RFID Applications

Passage C   RFID for Medical Application and Pallet Tracking

Passage A An Overview of RFID

I. An Introduction

Radio-frequency identification (RFID) is the wireless use of electromagnetic fields to transfer data, for the purposes of automatically identifying and tracking tags attached to objects. It is a generic term that is used to describe a system that transmits the identity (in the form of a unique serial number) of an object or person wirelessly, using radio waves. It is grouped under the broad category of automatic identification technologies.

RFID technology is used today in many applications, including security and access control, transportation and supply chain tracking. It is a technology that works well for collecting multiple pieces of data on items for tracking and counting purposes in a cooperative environment.

Typical RFID systems are made up of three components: readers (interrogators), antennas and tags (transponders) that carry the data on a microchip. The tags contain electronically stored information. Some tags are powered by electro magnetic inductionfrom magnetic fields produced near the reader. Some types collect energy from the interrogating radio waves and act as a passive transponder. Other types have a local power source such as a battery and may operate at hundreds of meters from the reader. Unlike a barcode, the tag does not necessarily need to be within line of sight of the reader and may be embedded in the tracked object. A typical RFID system comfiguration diagram is shown as Figure 3.1.

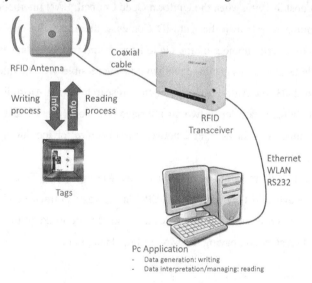

Figure 3.1 A typical RFID system configuration diagram

The primary benefits of RFID technology over standard barcode identification are:
- Information stored on the tag can be updated on demand
- Large data storage capacity (up to 4k bits)
- High read rates
- Ability to collect data from multiple tags at a time
- Data collection without line-of-sight requirements

- Longer read range
- Greater reliability in harsh environments
- Greater accuracy in data retrieval and reduced error rate

II. The History of RFID

In 1945, Léon Theremin invented an espionage tool for the Soviet Union which retransmitted incident radio waves with audio information. Even though this device was not an identification tag, it was likewise passive, being energized and activated by waves from an outside source.

The initial device was passive, powered by the interrogating signal, and was demonstrated in 1971 to the New York Port Authority and other potential users, and consisted of a transponder with 16 bit memory for use as a toll device.

An early demonstration of *reflected power* (modulated backscatter) RFID tags, both passive and semi-passive, was performed in 1973. The portable system operated at 915 MHz and used 12-bit tags. This technique is used by the majority of today's UHFID and microwave RFID tags.

Later, companies developed a low-frequency (125 kHz) system, featuring smaller transponders. A transponder encapsulated in glass could be injected under the cows' skin. This system is still used in cows around the world today.

In the early 1990s, IBM engineers developed and patented an ultra-high frequency (UHF) RFID system. UHF offered longer read range (up to 20 feet under good conditions) and faster data transfer.

UHF RFID got a boost in 1999, when the Uniform Code Council, EAN International, Procter & Gamble and Gillette put up funding to establish the Auto-ID Center at the Massachusetts Institute of Technology. Their idea was to put only a serial number on the tag to keep the price down. Data associated with the serial number on the tag would be stored in a database that would be accessible over the Internet.

Between 1999 and 2003, the Auto-ID Center opened research labs in Australia, the United Kingdom, Switzerland, Japan and China. It developed two air interface protocols (Class 1 and Class 0), the Electronic Product Code (EPC) numbering scheme, and a network architecture for looking up data associated on an RFID tag on the Internet.

Some of the biggest retailers in the world-Albertsons, Metro, Target, Tesco, Wal-Mart and the U.S. Department of Defense have said they plan to use EPC technology to track goods in their supply chain. The pharmaceutical, tire, defense and other industries are also moving to adopt the technology. EPC global ratified a second-generation standard, paving the way for broad adoption.

III. Readers

RFID systems can be classified by the type of tag and reader. A Reader is also called an interrogator. A typical reader is a device that has one or more antennas that emit radio waves and receive signals back from the tag. The reader then passes the information in digital form to a computer system. Readers can be configured with antennas in many formats including handheld devices, portals or conveyor mounted. There are 3 kinds of RFID system: PRAT, ARPT and ARAT.

A Passive Reader Active Tag (PRAT) system has a passive reader which only receives radio signals from active tags (battery operated, transmit only). The reception range of a PRAT system reader can be adjusted from 1-2,000 feet (0-600m), allowing flexibility in applications such as asset protection and supervision.

An Active Reader Passive Tag (ARPT) system has an active reader, which transmits interrogator signals and also receives authentication replies from passive tags.

An Active Reader Active Tag (ARAT) system uses active tags awoken with an interrogator signal from the active reader. A variation of this system could also use a Battery-Assisted Passive (BAP) tag which acts like a passive tag but has a small battery to power the tag's return reporting signal.

Fixed readers are set up to create a specific interrogation zone which can be tightly controlled. This allows a highly defined reading area for when tags go in and out of the interrogation zone. Mobile readers may be hand-held or mounted on carts or vehicles.

IV. Tags

An RFID tag is a microchip combined with an antenna in a compact package; the packaging is structured to allow the RFID tag to be attached to an object to be tracked. RFID tags contain at least two parts: an integrated circuit for storing and processing information, modulating and demodulating a radio-frequency(RF) signal, collecting DC power from the incident reader signal, and other specialized functions; and anantennafor receiving and transmitting the signal. The tag information is stored in a non-volatile memory. The RFID tag includes either fixed or programmable logic for processing the transmission and sensor data, respectively. RFID reader and tag diagram are shown as Figure 3.2.

Figure 3.2 RFID reader and tag diagram

RFID tags are categorized as either passive or active.

Passive tags do not have an integrated power source and are powered from the signal carried by the RFID reader. When radio waves from the reader are encountered by a passive RFID tag, the coiled antenna within the tag forms a magnetic field. The tag draws power from it, energizing the circuits in the tag. The tag then sends the information encoded in the tag's memory. However, to operate a passive tag, it must be illuminated with a power level roughly a thousand times stronger than for signal transmission. That makes a difference in interferences and in exposures to radiation.

An RFID tag is an active tag when it is equipped with a battery that can be used as a partial or complete source of power for the tag's circuitry and antenna. Some active tags contain replaceable batteries for years of use; others are sealed units. As a result of the built-in battery, active tags can operate at a greater distance and at higher data rates in return for limited life driven by the longevity of the built in battery and higher costs. For a lower cost of implementation, passive tags are more attractive solution.

Tags may either be read-only, having a factory-assigned serial number that is used as a key into a database, or may be read/write, where object-specific data can be written into the tag by the system user. Field programmable tags may be write-once, read-multiple; "blank" tags may be written with an electronic product code by the user.

V. Frequency

RFID is considered as a nonspecific short range device. It can use frequency bands without a license. Nevertheless, RFID has to be compliant with local regulations (ETSI, FCC etc.).

- LF: 125 kHz - 134.2 kHz : low frequencies;
- HF: 13.56 MHz : high frequencies;
- UHF: 860 MHz - 960 MHz : ultra high frequencies;
- SHF: 2.45 GHz : super high frequencies.

VI. New Words

electromagnetic [ɪ,lektrəʊmæg'netɪk]　　*adj.* 电磁的
interrogator [ɪn'terəʊgeɪtə]　　*n.* 应答器
antenna [æn'tenə]　　*n.* 天线
tag [tæg]　　*n.* [电子]标签
transponder [træns'pɒndə(r)]　　*n.* 转发器；转调器；变换器
microchip ['maɪkrəʊtʃɪp]　　*n.* 微型集成电路片，微芯片
interrogating [ɪn'terəgeitɪŋ]　　*v.* 询问
passive ['pæsɪv]　　*adj.* 无源的；被动的
barcode ['bɑrkod]　　*n.* 条形码；条码技术
espionage ['espɪɒnɑ:ʒ]　　*n.* 间谍；间谍活动；刺探
conveyor [kən'veɪə]　　*n.* 输送机，传送机；传送带
programmable ['prəʊgræməbl]　　*adj.* 可编程的；可设计的
passive ['pæsɪv]　　*adj.* 被动的
active ['æktɪv]　　*adj.* 主动的
radiation [,reɪdɪ'eɪʃn]　　*n.* 辐射；发光；放射物

VII. Phrases

tracking tags　　跟踪标签
radio waves　　无线电波
supply chain　　供应链；供给链；供需链
reflected power　　反射功率
Battery-Assisted Passive (BAP) tag　　蓄电池无源标签
sensor data　　传感器数据
replaceable batteries　　可替换电池
electromagnetic spectrum　　电磁波谱

VIII. Abbreviations

RFID　　Radio-frequency identification　　射频识别
UHFID　　Ultra High Frequency Identification　　特高频识别
UHF　　Ultra High Frequency　　特高频
PRAT　　Passive Reader Active Tag　　无源阅读器有源标签

ARPT	Active Reader Passive Tag	有源阅读器无源标签
ARAT	Active Reader Active Tag	有源阅读器有源标签
BAP	Battery-Assisted Passive	无源蓄电池
ETSI	European Telecommunications Standards Institute	欧洲电信标准协会
LF	Low Frequency	低频
HF	High Frequency	高频
SHF	Super High Frequency	超高频
FCC	Federal Communications Commission	联邦通信委员会

Passage B RFID Applications

I. Identification

Electronic passports like "e-passports" were adopted electrically after the 9/11 attack. After the terrible tragedy broke heart of United States, the American government became aware of the importance of checking VISAs and passports correctly. The USD department of State soon let people who wanted to enter US to use RFID tag embedded electronic passports instead of traditional barcode based passports. The European Union also endorsed the inclusion of biological information in e-passports. The EU Justice and Home Affairs Council decided to include fingerprints as a second mandatory identifier on passports in 2004. In addition, RFID can be used in e-ID cards in various countries. For example, in the United Kingdom, Prime Minister Tony Blair and his Labor Party convinced the nation to adopt biometrically-enhanced national identification cards. Tony Blair's administration announced its will to implement RFID tag embedded national identification card in late 2004. China is another case where the e-ID card is used today. As a matter of fact, China is the country where e-ID card is widely and largely adopted today. The Beijing Olympics held in 2008 lit the fuse of adoption. The largest smart card project was implemented as a part of preparing the most prominent international sports event. In 2008, the Chinese government supplied 1.2 billion dollars of RFID readers and 2.25 billion dollars of RFID embedded smart cards to citizens. This made China the world's largest market for RFID.

II. Transportation

Public transportation is another popular sector for RFID technology applications. RFID based electronic toll collection technology is one of the oldest and widespread RFID implementation. As soon as an RFID tag embedded car arrives at a toll booth, the RFID reader scans and reads the information that the RFID tag contains. The driver will pay debits according to the price that electronic reader suggests.

In the US, electronic toll collection is regarded as an efficient and effective method that eliminates long lines of traffic at a toll booth. RFID based toll collection is also adopted in criminal cases because it enables prosecutors to identify the exact location of the criminal's car. In Republic of Korea, the Korean government has set credit card-linked electronic toll collection system called "Hypass" especially for collecting transportation tolls on express ways.

If an RFID tag is embedded on their cars, drivers can pass the tollbooth without stopping the car because RFID reader scan the data immediately and handle the whole payment process in about 5s. HongKong SAR, China launched a similar public transportation toll collection system in 1997 and the "Octopus Card" is now internationally famous for its convenience. This system is able to handle 10 million transactions per day and includes all modes of public transport. Republic of Korea has set credit card-linked electronic toll collection system called "Hypass" especially for collecting transportation tolls on express ways. RFID technology is also implemented in railroad toll collection in India, where railroads are the most widely used form of public transportation. If an RFID tags embedded on their trains, drivers can pass the toll booth without stopping the car because RFID reader scan the data immediately and handle the whole payment process in about 5 s.

In addition, RFID has been used as a critical technology to promote efficiency and transparency for public transportation system in developing countries. RFID application in transportation is shown as Figure 3.3.

Figure 3.3 RFID application in transportation

III. Environmental Applications

RFID technology can be widely applied in environmental applications. Adopting an RFID system in waste management is the most prominent way of using RFID to ensure efficient, eco-friendly waste management among lots of countries in the world. PAYT program done by European Union (EU) is the pioneer of this field. PAYT is an RFID based waste pricing model that allows each individuals or each household to pay for the tag along with the total amount of waste they throw. Since each household and individual has a waste box in which RFID tags are embedded, the exact volume of waste can be calculated. In Europe this incentive based system has been proven to be a powerful policy tool for reducing the total amount of waste and for encouraging recycling. Similar systems are broadly implemented in the US. In South Korea, the ministry of environment introduced it to industry and urged them to use an RFID based waste management system, especially in medical waste management. RFID technology is implemented in waste management in developed and developing countries, but the purpose of adoption is somewhat different from Europe to the US. India, the second-most populated country in the world, has adopted RFID technology to cope with the rapid increase of volume and types of waste. Similarly in 2010, what China faced were the World Expo and huge amounts of construction waste that comprised 30% to 40% of the total urban waste. Shanghai was chosen for a pilot project using an RFID based waste management system. All the waste dumping trucks had an embed RFID tag and volume of waste they carry was checked by the local government. Another interesting case of environmental application emerges from Republic of Korea. The Republic of Korea government operates U-Street Trees Systems through which the exact location and status of street trees can be monitored. Information about location and status of street trees are collected by an RFID tag that is attached to each tree is saved in a web information system, so trees can be managed effectively. Kim et al. claim that this web based information system could manage information remotely with an interactive system.

IV. Healthcare and welfare

RFID enables hospitals to manage their equipment more easily and save expenses in public health areas. The US government agencies like FDA have also already used RFID tag in monitoring drug industry.

Since American hospitals handle almost 4,000 medicines per day, medication errors can easily occur. Even though it is not yet commercialized, an RFID identification system for the visually impaired people is being developed by engineers in Pakistan with the support of the Pakistani government.

Ⅴ. Defense and Security

The history of RFID technology was started from the need for ensuring national security. Almost 60 years have passed since US army developed RFID based identification system to identify allies and enemies, RFID technology is still used for protecting people. For instance, Weinstein and Konsysnki& Smith reported how the US Army and Navy implement RFID technology in cargo containers to identify materials. The US Army and Navy implement RFID not only identify US troops' own weapons and containers but also identify enemies in battle. RFID systems are also important in terms of airport and port security. After the 9/11 terrorist attack on the United States, President George W. Bush let all the airport and port in US adopt identification systems based on RFID technology to protect its nation from additional terrorist attacks. In 2012, China decided to implement an RFID based e-Seal system to increase security and efficiency. In addition, RFID technology can be used effectively in prison management 8 and child protection. In some countries like Japan and Republic of Korea, the RFID tag is implemented in child protection monitoring. Fingerprint recognition phone is shown as Figure 3.4.

Figure 3.4　Fingerprint recognition phone

Ⅵ. New Words

barcode ['bɑrkod]	n. 条形码；条码技术
fingerprint ['fɪŋɡɚprɪnt]	n. 指纹；手印
remotely [rɪ'məutlɪ]	adv. 远程地

Ⅶ. Phrases

biological information	生物信息
Octopus Card	八达通卡
waste management	废物管理
PAYT program	垃圾按量收费项目
waste dumping trucks	垃圾堆卸载卡车
embed RFID tag	嵌入式射频识别标签

Passage C RFID for Medical Application and Pallet Tracking

Radio Frequency Identification (RFID) systems have been successfully applied and shown its worth in many fields of manufacturing, transportation, agriculture, healthcare and supply chain, just to name a few. The use of RFID in healthcare and hospital services has increased significantly over the years due to its credibility and accuracy. RFID technology is acting crucial in item level tracing and tracking systems in hospital scenarios. The aims of those applications are to express and display the role of RFID technology in maintaining the inventory level of medicines and surgical items in hospital warehouses and increase its impact in healthcare. One of the crucial points that should be noted in healthcare is that the medicines and surgical items should be made available at any point of time. The implementation of RFID technology in this area is to ensure patient safety and satisfaction by maintaining a perfect inventory of medicines and surgical items.

I. RFID for Patients Monitoring Applications

This application is presented about wireless sensor networks (WSN) that can observe the human physiological signals by ZigBee, which is provided with lower power consumption, small volume, high expansion, stylization and two way transmission, etc. This application developed a set of hospital tracking sensor network system by ZigBee's characteristics, which is embedded sensors, such as the biosensor to observe the body temperature. The biosensor transmits measured signals via ZigBee and then sends to remote wireless monitor for acquiring the observed human physiological signals. The remote wireless monitor is constructed of ZigBee and personal computer (PC). The measured signals send to the PC, which can be data collection. When the measured signals over the standard value, the personal computer sends Global System for Mobile Communication (GSM) short message to the doctor. The doctor can use the PC or mobile to observe the observed human physiological signals in the remote place.

Implementation of RFID technology for patient's monitoring will be resulting this technology to be great demand in health care sectors. Hospital is in hectic environment where countless number of people and items move and interact with one another every day.

Patients waiting time can be increased due to some simple staff such as in finding staff member and equipment. Inefficient patient flow in emergency rooms, in-patient beds, out-patient exam rooms, and procedure areas, can give a bad impact on the entire facility. There are many different areas that could quickly benefit if only people understood what is happening and could get access to timely information.

Healthcare organizations can improve their operating system, by providing an easy way to collect data and by delivering a convenient way to make the information easy to use. In the case of emergency and dangerous situations the alert system to the doctor is needed immediately. Thus, RFID Tracking System,which promote doctor to patient communication and indicate the status of the patient in the hospital to doctor through SMS, is proposed. This way of communication can be done by integrating RFID Tracking

System with the GSM technology. To start to implement the system, each patient will be given a tag to be monitored. In the case of emergency, the doctors will get message through handphone and can immediately attend the patient without any failure. This is convenient process to monitor the patient's health conditions from any distance. By using GSM technology, the user is able to do real time communication in an indoor environment as well as for longer coverage. Main Components of RFID based tracking system are shown as Figure 3.5.

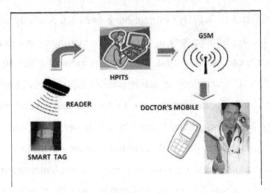

Figure 3.5　Main components of RFID based Tracking System

II. RFID-based Hospital Real-time Patient Management System

Healthcare providers (i.e., hospitals) traditionally use a paper-based "flow chart" to capture patient's information during registration time, which is updated by the on duty nurse and handed over to the incoming staff at the end of each shift. Although, the nurses spent large amount of time on updating the paper work at the bedside of the patient, it is not always accurate, because this is handwritten.

The nurses play a vital role at the hospital system in the success of both inpatient and outpatient care. They also play a very important role in bridging to execute clinical orders or to communicate information between the hospital and the patient that motivates to evaluate the potential of RFID technology, and to reinforce the critical job of information handover.

RFID is one of the emerging technologies offering a solution, which can facilitate automating and streamlining safe and accurate patient identification, tracking, and processing important health related information in health care sector such as hospitals.

Each RFID tag/wristband is identified by a Unique Identification Number (UIN) that can be programmed either automatically or manually and then password protected to ensure high security.

RFID wristband can be issued to every patient at registration, and then it can be used to identify patients during the entire hospitalization period. It can also be used to store patient important data (such as name, patient ID, drug allergies, drugs that the patient is on today, blood group, and so on) in order to dynamically inform staff before critical. RFID encoded wrist band data can be read through bed linens while patients are sleeping without disturbing them.

RFID technology provides a method to transmit and receive data from a patient to health service provider medical professionals without human intervention (i.e., wireless communication). It is an automated data-capture technology that can be used to identify, track, and store patient information electronically contained on RFID wristband (i.e., smart tag). Although, medical professionals/consultants can access/

update patient's record remotely via Wi-Fi connection using mobile devices such as PDA (Personal Digital Assistant), laptops and other mobile devices. Wi-Fi (wireless fidelity), or wireless local area networks (WLAN) that allow healthcare provider (e.g., hospitals) to deploy a network more quickly, at a lower cost, and with greater flexibility than a wired system.

RFID technology mainly consists of a Transponder (smart tag), a Reader and Healthcare Provider IT Systems (HPITS) as shown in Figure 3.6. Each tag attached to the patient wristband contains an antenna and a tiny microchip smaller than a grain of sand.

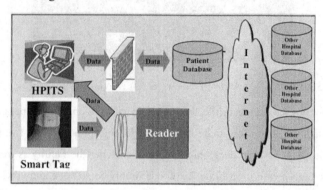

Figure 3.6 Main components of RFID-based Patient Management System

The antenna picks up radio-waves or electromagnetic energy beamed at it from a reader device and enables the chip to transmit patient's unique ID to the reader device, allowing the patient to be remotely identified. The reader converts the radio waves reflected back from the patient wristband into digital information that can then be passed onto HPITS for processing.

Patient's basic important data (e.g., patient ID, name, age, location, drug allergies, blood group, drugs that the patient is on today) can be stored in the patient's back-end databases for processing. Patient databases can also be linked through Internet into other hospitals' databases.

III. RFID for Pallet Tracking

Many companies lease pallets and other logistics containers in order to offload the cost of maintaining their own pallet fleet. However, managing the return of these items has often proven challenging for pallet providers. Although many companies currently use cheaper wood pallets, demand for reusable plastic pallets has increased in recent years because of fears about cross contamination from meat and other food products shipped on wooden pallets (which are often scrapped afterward), and new International Plant Protection Convention (IPPC) requirements for the fumigation of wood pallets to prevent the spread of invasive insects. Plastic pallets also last longer and are lighter than their wood counterparts, which can save on freight costs.

SvenskaRetursystem (SRS) distributes more than one million plastic pallets throughout Sweden. Keeping track of each pallet is a formidable, but essential, task. Deposits are required to cover costs should pallets be damaged or lost, and SRS was constantly running into problems with pallets being returned later than expected, or not at all. The outdated inventory system wasn't adequately keeping track of where pallets were being lost or damaged, resulting in SRS incurring the costs. The Radio – Frequency Identification (RFID) tags applied to opposite sides of the pallet are shown as Figure 3.7.

Figure 3.7 The Radio-Frequency Identification (RFID) Tags applied to opposite sides of the pallet

A system for tracking the pallets while they were in circulation was desperately needed. Bar code technology was proposed, but determined to be insufficient for the application for two reasons. First, because these pallets are for carrying fresh produce, each pallet re-entering the facility is treated with a variety of cleaning chemicals and high-pressure washing. Bar codes simply could not endure these conditions. Also, the life of any pallet inevitably involves a significant amount of bumping and bruising, and even a small scratch can render a barcode useless.

To provide real-time visibility of pallets in circulation, each pallet was outfitted with a ruggedized tag that would remain with it for its entire lifetime. Instead of mounting these tags to the exterior of the pallet (as is necessary with bar codes due to the line-of-sight requirement), engineers at pallet manufacturer Arca System, Perstorp, Sweden, embedded the tags in the interior structure of the two-piece pallet. Two tags are embedded inside opposite corners of the pallets to ensure that regardless of orientation on the conveyer, one tag will be picked up. Once installed, the tags can be written to with information and read by the antenna. When the pallets complete a cycle, the tags are cleared and rewritten.

Two plate antennas at the entrance to the pallet-cleaning machine read the tags. Pallet history is downloaded for each pallet, and if damage is evident, the proper customer is charged. Likewise, the pallets whose origin is unknown can be identified by reading the tags.

In this application, pallets leave SRS and arrive at the warehousing hub of a major supermarket. Supermarket operators use a portable handheld scanner to read and write time and date stamp, as well as handling instructions, to the tag. The pallets are then loaded with perishable goods, and the tags are written to with information such as product expiration date and storage instructions. The full pallets are then delivered to the neighborhood grocery store, where grocery workers scan them as they arrive to ensure freshness and quality. Empty pallets are loaded into the trucks and returned to SRS where they are again read. After relevant charges are processed, pallet information is then cleared, and the tag is written to with date/time information for its next voyage out into the world. Results of the RFID system are impressive. Each pallet has a self-contained recordkeeping system, eliminating the paperwork that is common with pallet shipments. The combination of resilient RFID tags and reusable pallets is a cost-effective solution: wooden pallets and bar-code labels are both consumables, while plastic pallets and RFID tags are recycled and reused. By accessing the shipment history of each incoming pallet, SRS is able to recoup for lost and damaged pallets by having record of who was responsible for the product at any given time. Customers pay for the length of time they have the pallets as well, so usage costs can be immediately determined and charged. Theft, formerly a rampant problem, is greatly reduced with RFID. If a stolen pallet reenters the facility, operators can determine when and where the pallet was taken.

IV. New Words

manufacturing [mænjʊ'fæktʃərɪŋ]	n. 制造业；工业	v. 制造
transportation [trænspɔr'teʃən]	n. 运输；运输系统	
stylization [staɪlɪ'zeʃən]	n. 格式化；仿效	
ZigBee	n. 无线个域网	
cost-effective [kɔ:st I'fektɪv]	adj. 划算的；成本效益好的	
bar-code ['bɑ:kəud]	n. 条形码	

V. Phrases

high expansion	高扩展
hand phone	手机
smart tag	智能标签
outdated inventory system	过时的库存系统
re-enter	重新进入
high-pressure	高压的；高气压的
line-of-sight	视线；视线角
recordkeeping system	记录系统

VI. Abbreviations

GSM	Global System for Mobile	全球移动通信系统
SMS	Short Messaging Service	短信服务
UIN	Unique Identification Number tong	通用国际互联网号码
HPITS	Healthcare Provider IT Systems	医疗保健信息技术系统
IPPC	International Plant Protection Convention	国际植物保护公约

Exercises

Translate the following sentences into Chinese or English.

1. Basically, RFID facilitates the control on a wide variety of processes in different stages of the lifecycle of a building, from its conception to its inhabitance.

2. In the food industry, it is important to trace the history and the localization of products aiming to guarantee quality and security in the food chain. An RFID system is proposed to determine the product provenance in the meat production industry. Furthermore, in consumer packaged goods management, several important companies require their providers to install RFID tags in pallets or boxes in order to improve the processes of storage, inventory and security.

3. RFID可以大大提高物流行业、制造业和公共信息行业的管理和操作效率。

4. RFID是一种非接触式的自动识别技术,通过射频信号自动识别目标对象。

参考译文　Passage A　RFID概述

I. 概述

射频识别（RFID）通过无线电磁场传输数据，自动识别并跟踪特定目标的标签。它是一个通用术语，用来描述采用无线电波对物体和人的身份进行无线传输的系统。这是宏观分组下的自动识别技术。

如今，RFID技术已被运用到多个领域，包括安全和访问控制、运输和供应链跟踪。这种技术适用于收集多个数据片段，从而达到在协同环境中跟踪和计数的目的。

典型的RFID系统由3个组成部分：阅读器（询问器）、天线，以及用微芯片携带数据的标签（也叫转发器）。标签包含电子化存储信息。一些标签是由磁场附近的阅读器产生的电磁感应驱动；另一些从交互无线电波中收集能量并作为无源转发器；还有一些有本地电源例如电池，可以通过数百米外的阅读器进行操作。与条形码不同，标签不一定要在阅读器的视线范围内，可以嵌入跟踪目标。

RFID技术在标准条形码识别下的主要优势有：

- 存储在标签上的信息可按需更新；
- 大容量存储数据（最多4k字节）；
- 阅读率高；
- 能同时从多个标签收集数据；
- 数据收集无需考虑视距要求；
- 有更大的阅读范围；
- 在恶劣环境下可靠性更强；
- 数据检索准确率高，错误少。

II. 射频识别简史

1945年，里昂·泰勒明为苏联发明了间谍工具，用于转播带有音频信息的无线电波。虽然这个装置不是识别标签，但它同样是可被外来电波激活并通电的无源装置。

最初的装置是无源的，由交互信号驱动。1971年，纽约港务局和其他一些潜在用户一起证实了该装置的可行性，该装置由一个16位内存的转发器组成，其中转发器作为数字设备。

1973年，具有反射功率的无源和半无源标签（调制后反向散射）RFID标签研制完成。便携式系统运行在915MHz而且使用12位标签。如今，这项技术被大多数UHFID和微波射频识别标签使用。

之后，几家公司开发了一个表面上类似于小型转发器的低频系统（125kHz）。该转发器可以被密封在玻璃中，注射到牛皮里。该系统仍适用于世界各地的牛群。

早在20世纪90年代，IBM工程师就开发并获得了特高频（UHF）RFID系统的专利。UHF可以满足更长的阅读距离（在良好的条件下可以高达20英尺），而且加快了传输速度。

UHF RFID在1999年得到了进一步发展，统一编码委员会、国际物品编码协会、宝洁和吉列公司一起投资在麻省理工学院建立了一个自动识别研究中心。他们的想法是在标签上加用序列号来降低价格。与标签上的序列号相联系的数据会保存于可通过因特网访问的数据库中。

在1999年至2003年间，自动识别中心在澳大利亚、英国、瑞士、日本和中国都开设了研究室，并开发出了两类空中接口协议（类1和类0）、电子产品代码（EPC）编码方案，以及网络体系结构，用于在因特网上查找与RFID标签相关的数据。

一些大型零售商——艾伯森超市、麦德龙超市、塔吉特公司、特易购、沃尔玛超市和美国国防部都宣称他们计划采用电子产品编码技术在供应链中追踪商品。制药业、轮胎制造业、国防部和其他行业也都转而采用这种技术。电子产品编码又批准了第二代标准，为广泛采用此技术铺平了道路。

Ⅲ. 阅读器

RFID系统可以分为标签和阅读器两种类型。

阅读器也被称为查询器。典型的阅读器装置有一根或多根天线发射无线电波并接收电子标签发送回来的信号。然后，阅读器以数字化形式将信息传送到计算机系统。阅读器可以根据天线配置成多种形式，包括手持设备，门户网站或输送机安装。

3种RFID系统：无源阅读器有源标签、有源阅读器无源标签和有源阅读器有源标签。无源阅读器有源标签（PRAT）系统有一个无源阅读器，只能从有源标签（电池供电的，仅供传输）接收无线信号。PRAT系统阅读器的接收范围涵盖1英尺调整至2000英尺（0~600m），可以灵活应用，如资产保护和监管。

有源阅读器无源标签（ARPT）系统有一个有源阅读器传送询问器信号，也接收无源标签反馈的认证信号。

有源阅读器有源标签（ARAT）系统采用由有源阅读器询问信号唤醒的有源标签。此类系统同样可以使用蓄电池无源（BAP）标签，就像无源标签一样运作，但有一个小电池为标签反馈报告信号提供电源。

设置固定阅读器用于创建一个特定的询问区，这个询问区可以被严格控制。这使标签在进出询问区时有一个高度明确的阅读区域。便携式阅读器可手持或安装在手推车或汽车上。

Ⅳ. 标签

RFID标签是由一个微型集成电路芯片和一根天线组成的微型器件，该器件的结构设计是为了让RFID标签可以被贴在物品上对这些物品进行跟踪。

RFID标签至少包括两个部分：一个是集成电路，用于存储和处理信息，调制和解调射频信号，从入射阅读信号中收集直流电，还有一些其他的专业性功能；另一个是天线，用于接收和发射信号。标签信息存储在非易失性储存器中。RFID标签包括固定的或可编程部分，分别用于处理传输和传感器的数据。

RFID标签被分为无源标签和有源标签。

无源标签没有集成的电源，而是从RFID阅读器提供的信号中获得。当阅读器中的无线电波遇到无源RFID标签时，标签内的线圈天线就会形成一种磁场。标签的能源来自于这个磁场，从而激活标签内的电路。然后，标签会发送编码在标签内存中的信息。然而，运行一个无源标签，相比用于信号传输能量，启动它需要多出上千倍的能量。这引起的信号干扰和辐射是不同的。

若RFID标签装有电池后，这个电池可作为局部的或完整的电源提供给标签的电路系统和天线，这时RFID标签是有源标签。一些有源标签装有可替换的电池可以使用好几年；还有一些是密封的。

由于内置电池，有源标签可以在更远的距离和更高的数据速率下运作，取代了内置电池有限的寿命和较高的成本。若要实现较低的费用，无源标签是一个更具吸引力的解决方案。

标签要么是只读的，带有生成指定序列号作为进入数据库的密钥。或者是可读/写的，系统用户可将特定目标数据写入标签。现场可编程的标签是一次写入，多次读出；用户可以用电子产品代码编写"空白"标签。

Ⅴ. 频率

RFID是一个非特定的短程设备，不需要许可就可以使用频段。但是，RFID必须符合本地管理规定（ETSI，FCC等）。

LF：125 kHz - 134.2 kHz：低频；

HF：13.56 MHz：高频；

UHF：860 MHz ~ 960 MHz：特高频；

SHF：2.45 GHz：超高频。

参考答案

1. 射频识别对建筑周期的每个阶段都有促进作用，从起初的概念到最后的实实在在的住房。

2. 在食品工业中，最重要的是查找产品的发展史和产地以保证食物链的质量和安全。（在第五部分中，作者会谈到射频识别系统如何将新鲜食物进行低温运输。此处原英文中未提及）有人建议射频识别系统可用于确定肉类行业中产品的来源。另外，对于一些高级打包商品的管理，一些重要的公司都要求他们的供货商在食品托盘或盒子上贴上RFID的标签，以确保储存的质量和安全。

3. The RFID can substantially improve the management and operation efficiency when used in industries of logistics, manufacture and public information service.

4. The Radio Frequency Identification (RFID) technology is a type of non-contact automatic identification technology realized through radio frequency communication.

科技文翻译技巧一：分句与合句

分句与合句，是冲破原文句法结构的束缚，翻译出规范译文的两个相辅相成的重要方法。所谓分句法，就是把原文某句话的成分加以分解，译成两个或两个以上的句子；而合句法，就是把原文两个或两个以上的句子加以合并，译成一个句子。

英语重形合，展现出"分岔式"的结构特征，而汉语重意合，展现出"排调式"的结构特征。正因为这个缘故，英语句子一般要长于汉语句子。特别是一些又长又复杂的句子，如果进行"字对字"的理解，往往会使译文读者感到费解。因此，翻译长句时，译者经常要采取"分而治之"的办法，以求行文更加顺畅。例如：

(1) This kind of two-electrodes tube consists of a tungsten filament, which gives off electrons when it is heated, and a plate toward which the electrons migrate when the field is in the right direction.

译文：这种二极管由一根钨丝和一个极板组成。钨丝受热时放出电子，当电场方向为正时，电子就移向极板。

分析：把两个which引导的定语从句从原句中分出，拆成两个独立分句，更符合汉语的表达习惯，意思明确，通俗易懂。

(2) The Data Link Layer provides the functional and procedural means to transfer data between network entities and to detect and possibly correct errors that may occur in the Physical Layer.

译文：数据链路层提供在网络实体之间传输数据的功能和方法，检查并尽量改正物理层可能出现的错误。

合句法多运用于英语简单句的翻译，特别是两个或两个以上的英文句子共用相同主语的时候，汉语一般不喜欢重复某一名词，或使用某一代词做主语，在这种情况下，可以将原句中的某些部分合并翻译，以使原文简洁通顺。例如：

(3) It is common practice that electric wires are made from copper.

译文：电线通常是铜制的。

分析：把主句和从句合二为一，避免了冗长复杂的表达方式。

(4) Energy is the scarcest resource of WSN nodes, and it determines the lifetime of WSNs.

译文：能量是无线传感网节点最稀缺的资源，决定了无线传感网的寿命。

Unit 4

Sensor

Passage A   Sensor Characteristics and Measurement Systems

Passage B   Capacitive Sensor Working Principle

Passage C   Sensor Networks Application in Agriculture and Forest Fire Prevention

Passage A Sensor Characteristics and Measurement Systems

I. Sensor Characteristics

Sensors provide an output signal in response to a physical, chemical, or biological measurand and typically require an electrical input to operate; therefore, they tend to be characterized in sensor specifications (commonly called "specs") or datasheets in much the same way as electronic devices. To truly understand sensors, and how sensors that measure the same measurand can differ, it is necessary to understand sensor performance characteristics. Unfortunately, there is no standard definition for many of these characteristics and different parts of the sensor community have different names for these characteristics, depending on the sensing domain. This confusion is compounded by manufacturers publishing an abundance of performance characteristics, which make it even more difficult for potential users to identify the ones relevant to their applications, and how they should be applied. The following section will describe these characteristics, using their most common names, and will reference alternative names where relevant.

1. Statistical Characteristics

(1) Sensitivity

Sensitivity is the change in input required to generate a unit change in output. If the sensor's response is linear, sensitivity will be constant over the range of the sensor and is equal to the slope of the straight-line plot (as shown in Figure 4.1). An ideal sensor will have significant and constant sensitivity. If the sensor's response is non-linear, sensitivity will vary over the sensor range and can be found by calculating the derivative of S with respect to x (dS/dx).

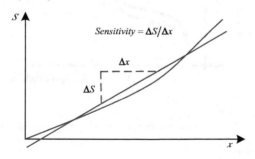

Figure 4.1 Sensor sensitivity

(2) Linearity

The linearity of the transducer is an expression of the extent to which the actual measured curve of a sensor departs from the ideal curve. Figure 4.2 shows a somewhat exaggerated relationship between the ideal, or least squares fit line and the actual measured or calibration line (Note in most cases, the static curve is used to determine linearity, and this may deviate somewhat from a dynamic linearity). Linearity is often specified in terms of percentage of nonlinearity, which is defined as:

$$\text{Nonlinearity }(\%) = \frac{D_{\text{in(max)}}}{\text{IN}_{f,s}} \times 100$$

Where Nonlinearity (%) is the percentage of nonlinearity, $D_{\text{in(max)}}$ is the maximum input deviation, $\text{IN}_{f,s}$ is the maximum, full-scale input.

The static nonlinearity defined by Equation is often subject to environmental factors, including temperature, vibration, acoustic noise level, and humidity. It is important to know under what conditions the specification is valid and departures from those conditions may not yield linear changes of linearity.

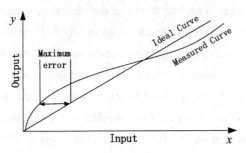

Figure 4.2　Ideal versus measured curve showing linearity error

(3) Hysteresis

Hysteresis is the difference between output readings for the same measurand, depending on the trajectory followed by the sensor. Hysteresis may cause false and inaccurate readings. Figure 4.3 represents the relation between output and input of a system with hysteresis. As can be seen, depending on whether path 1 or 2 is taken, two different outputs, for the same input, can be displayed by the sensing system.

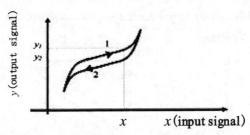

Figure 4.3　An example of a hysteresis curve

2. Dynamic Characteristics

Dynamic characteristics tell us about how well a sensor responds to changes in its input. For dynamic signals, the sensor or the measurement system must be able to respond fast enough to keep up with the input signals. Dynamic Characteristice is shown as Figure 4.4.

(1) Step Response

Speed of response: indicates how fast the sensor (measurement system) reacts to changes in the input variable. (Step input) Rise time: the length of time it takes the output to reach 90% of full response when a step is applied to the input. Time constant: (1st order system) the time for the output to change by 63.2% of its maximum possible change. Settling time: the time it takes from the application of the input step until the output has settled within a specific band of the final value. Transfer Function: a simple, concise and complete way of describing the sensor or system performance $H(s) = Y(s)/X(s)$ where $Y(s)$ and $X(s)$ are the Laplace Transforms of the output and input respectively. Sometimes, the transfer function is displayed graphically as magnitude and phase plots VS frequency (Bode plot).

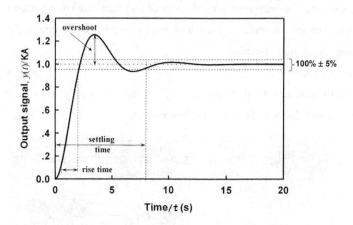

Figure 4.4 Dynamic Characteristics

(2) Frequency Response

Frequency Response describes how the ratio of output and input changes with the input frequency (sinusoidal input).

Dynamic error, $\delta(\omega) = M(\omega) - 1$ a measure of the inability of a system or sensor to adequately reconstruct the amplitude of the input for a particular frequency.

Bandwidth the frequency band over which $M(\omega) \geqslant 0.707$ (−3 dB in decibel unit).

Cutoff frequency: the frequency at which the system response has fallen to 0.707 (−3 dB) of the stable low frequency.

II. Measurement Systems

(1) Primary Sensing Element

This is the element that first receives energy from the measured medium and produces an output depending in some way on the measured quantity. The output is some physical variable, e.g. displacement or voltage. An instrument always extracts some energy from the measured medium. The measured quantity is always disturbed by the act of measurement, which makes a perfect measurement theoretically impossible. Good instruments are designed to minimize this loading effect.

(2) Variable-Conversion Element

It may be necessary to convert the output signal of the primary sensing element to another more suitable variable while preserving the information content of the original signal .This element performs this function.

(3) Variable Manipulation Element

An instrument may require that a signal represented by some physical be mainpulated in some way. By manipulation we mean specifically a change in numerical value according to some define rule but a preservation of the physical nature of the variable. This element performs such a function.

(4) Data-Transmission Element

When functional elements of an instrument are actually physically separated, it becomes necessary to transmit the data from one to another. This element performs this function.

(5) Data-Presentation Element

If the information about the measured quantity is to be communicated to a human being for monitoring, controlling, or analysis purpose, it must be put into a form recognizable by one of the human senses. This element performs this "translation" function.

(6) Data-Storage Element

Some applications require a distinct data storage which can easily recreate the stored data upon command. The Measurement Systems is shown as Figure 4.5.

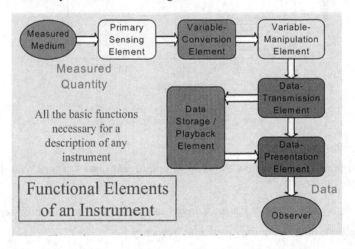

Figure 4.5　Measurement Systems

Ⅲ. **New Words**

sensors ['sensəz]	n. 传感器，灵敏元件
sensitivity [ˌsensɪ'tɪvɪti]	n. 灵敏性，敏感，感受性
slope [sloʊp]	n. 斜率
derivative [dɪ'rɪvətɪv]	n. 导数，微商
linearity [ˌlɪnɪ'ærətɪ]	n. 线性，直线性
transducer [trænz'duːsər]	n. 传感器，变频器，变换器
vibration [vaɪ'breʃən]	n. 震动；摆动；感受
humidity [hjuː'mɪdɪti]	n. 湿度；潮湿
hysteresis [ˌhɪstə'riːsɪs]	n. 滞后作用，磁滞现象；滞变
trajectory [trə'dʒektəri]	n. [物]弹道，轨道；[几]轨线
magnitude ['mægnɪtuːd]	n. 量级；巨大，广大

Ⅳ. **Phrases**

sensor community	传感社区
step response	阶跃响应
dynamic characteristics	动态特性

settling time	调节时间
transfer function	传递函数
phase plots	相位图
frequency response	频率响应
dynamic error	动态误差
cutoff frequency	截止频率
primary sensing element	初级感测元件
loading effect	负载效应
variable-conversion element	可调变换元件
data-transmission element	数据传输元件
data-presentation element	数据显示元件
data storage element	数据存储元件

Passage B Capacitive Sensor Working Principle

Capacitive sensing is becoming a popular technology to replace optical detection methods and mechanical designs for applications like proximity/gesture detection, material analysis, and liquid level sensing. The main advantages that capacitive sensing has over other detection approaches are that it can sense different kinds of materials (skin, plastic, metal, liquid), it is contactless and wear-free, it has the ability to sense up to a large distance with small sensor sizes, the PCB sensor's cost is low, and it is a low power solution.

I. Capacitance Measurement Basics

Capacitance is the ability of a capacitor to store an electrical charge. A common form – a parallel plate capacitor – the capacitance is calculated by $C = Q / V$, where C is the capacitance related by the stored charge Q at a given voltage V. The capacitance (measured in Farads) of a parallel plate capacitor (as shown in Figure 4.6) consists of two conductor plates and is calculated by:

$$C = \frac{\varepsilon_r \times \varepsilon_0 \times A}{d}$$

Equation is characterized by:

- A is the area of the two plates (in meters)
- ε_r is the dielectric constant of the material between the plates
- ε_0 is the permittivity of free space (8.85×10^{-12} F/m)
- d is the separation between the plates (in meters)

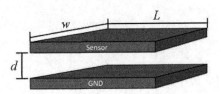

Figure 4.6 Parallel Plate Capacitor

The plates of a charged parallel plate capacitor carry equal but opposite charge spread evenly over the surfaces of the plates. The electric field lines start from the higher voltage potential charged plate and end at the lower voltage potential charged plate. The parallel plate equation ignores the fringing effect due to the complexity of modeling the behavior but is a good approximation if the distance between the plates is small compared to the other dimensions of the plates so the field in the capacitor over most of its area is uniform. The fringing effect occurs near the edges of the plates, and depending on the application, can affect the accuracy of measurements from the system. The density of the field lines in the fringe region is less than directly underneath the plates since the field strength is proportional to the density of the equipotential lines. This results in weaker field strength in the fringe region and a much smaller contribution to the total measured capacitance. Figure 4.7 displays the electric fields lines path of a parallel plate capacitor.

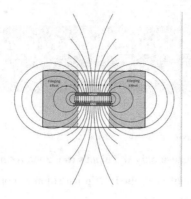

Figure 4.7　Electric Fields of a Parallel Plate Capacitor

II. Capacitive Sensing - How it Works

Capacitive sensing is a technology based on capacitive coupling that takes the capacitance produced by the human body as the input. It allows a more reliable solution for applications to measure liquid levels, material composition, mechanical buttons, and human-to-machine interfaces. A basic capacitive sensor is anything metal or a conductor and detects anything that is conductive or has a dielectric constant different from air. Figure 4.8 displays three basic implementations for capacitive sensing: proximity/gesture recognition, liquid level sensing, and material analysis. More details about the sensor topology can be found in the Capacitive Sensor Topologies section.

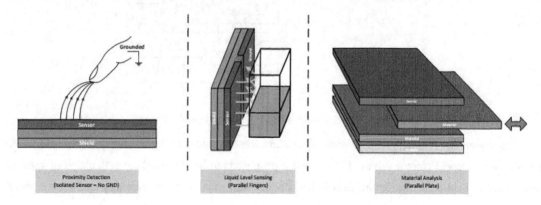

Figure 4.8　Basic Implementions for Capacitive Sensing

III. New Words

underneath [ʌndə'niːθ]　　　　　　　n. 下面；底部　adv. 在下面；在底下

IV. Phrases

optical detection methods　　　光学检测方法
capacitive sensing　　　　　　电容感应
capacitance Measurement Basics　电容测量基础
parallel Plate Capacitor　　　　平行板电容器
fringe region　　　　　　　　边缘区
capacitive coupling　　　　　　电容耦合

Passage C Sensor Networks Application in Agriculture and Forest Fire Prevention

Ⅰ. Application of Sensor Networks in Agriculture

Drip irrigation saves water because only the plant's root zone receives moisture. Little water is lost to deep percolation if the proper amount is applied. Drip irrigation is popular because it can increase yields and decrease both water requirements and labor. Drip irrigation requires about half of the water needed by sprinkler or surface irrigation. Lower operating pressures and flow rates result in reduced energy costs. A higher degree of water control is attainable. Plants can be supplied with more precise amounts of water. Disease and insect damage is reduced because plant foliage stays dry. Operating cost is usually reduced. Federations may continue during the irrigation process because rows between plants remain dry. The application of smart irrigation to filed is shown as Figure 4.9.

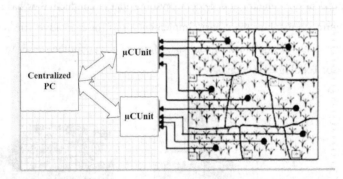

Figure 4.9 Application of Smart Irrigation to Field

The automated control system consists of moisture sensors, analog to digital converter, microcontroller, relay driver, solenoid control valves. The important parameters to be measured for automation of irrigation system are soil moisture. The entire field is first divided into small sections such that each section should contain one moisture sensor.

These sensors are buried in the ground at required depth. Once the soil has reached desired moisture level the sensors send a signal to the micro controller to turn on the relays, which control the valves.

Soil moisture sensors are designed to estimate soil volumetric water content based on the dielectric constant (soil bulk permittivity) of the soil. The dielectric constant can be thought of as the soil's ability to transmit electricity. The dielectric constant of soil increases as the water content of the soil increases. This response is due to the fact that the dielectric constant of water is much larger than the other soil components, including air. Thus, measurement of the dielectric constant gives a predictable estimation of water content.

Bypass type soil moisture irrigation controllers use water content information from the sensor to either allow or bypass scheduled irrigation cycles on the irrigation timer. The microcontroller has an adjustable threshold setting and, if the soil water content exceeds that setting, the event is bypassed. The soil water content threshold is set by the user.

The required readings can be transferred to the Remote Computer via ZigBee for further analytical studies, through the serial port present on microcontroller unit. While applying the automation on large fields more than one such microcontroller units can be interfaced to the Centralized Computer the microcontroller unit has in-built timer in it, which operates parallel to sensor system. In the case of sensor failure the timer turns off the valves after a threshold level of time, which may prevent the further disaster. The microcontroller unit may warn the pump failure or insufficient amount of water input with the help of flow meter.

The microcontroller based irrigation system proves to be a real time feedback control system which monitors and controls all the activities of drip irrigation system efficiently.

The present proposal is a model to modernize the agriculture industries on a small scale with optimum expenditure. Using this system, one can save manpower, water to improve production and ultimately profit. The real time automated irrigation system is shown as Figure 4.10.

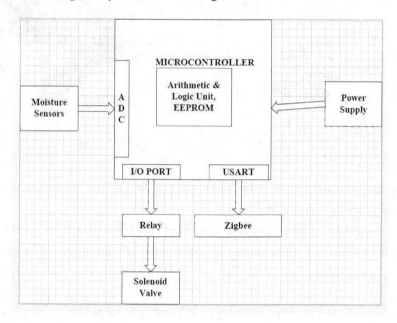

Figure 4.10 Real Time Automated Irrigation System

II. Application of Sensor Networks in Forest Fire Prevention

The application of this work focuses on a sensor network used to detect forest fire hazards. These events are related to different variables that exist in Earth's atmosphere. In particular, temperature, atmospheric pressure, water vapor and the gradients and interactions of each of these variables play an important role in the occurrence of these phenomena, as well as their evolution over time.

1. System Design

A diagram of the complete system can be seen in Figure 4.11. The prototype design consists of a sensor network that retrieves the necessary data, transmits it wirelessly and stores it in a computer for later analysis.

The sensors used measure temperature, relative and absolute humidity, wind speed and direction, and rainfall. These sensors send the data by radio frequency to a station where they are stored. A master node requests the data to the station and then packages it, together with data from other sensor nodes arranged in the bark of trees taken as sample. The master node also reads through an analog to digital converter, the voltage in the battery used as power supply and the intensity of current delivered by the solar panel of the system. Then the master node is responsible for transmitting all the data obtained through the cellular network, to a remote computer. That computer analyzes the data, calculates the risk of fires and presents data on a web server. Figure 4.12 shows a system prototype of the master node, which was used to make the first tests.

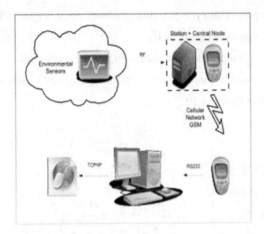

Figure 4.11 Complete System   Figure 4.12 Testing Prototype

2. Data Recovery

Within any system whose purpose is the analysis of data for further consideration, one of the central elements to consider are the sensors used.

The meteorological data required for calculations in relation to the Hazard Fire Meteorological Index, are those that affect the ease with which fires are started, its speed of propagation, and the effect of the fire in the environment. In this particular case, we used temperature sensors, humidity, soil moisture, atmospheric pressure, wind, and precipitations. Sensors were used from various commercial companies (PC Weather Station and Vernier), however, we decided to design our own versions of the sensors. In this way, the sensors used to measure soil moisture and humidity inside the bark of a tree are own versions, while the rest of the sensors will continue to use the commercial versions.

Figure 4.13 shows the sensors used to measure temperature, humidity, atmospheric pressure, wind speed and direction, and rainfall.

Figure 4.13 Sensors Used

Ⅲ. **New Words**

moisture ['mɒɪstʃə(r)]	n. 水分；湿度；潮湿；降雨量
percolation [ˌpɜːkə'leʃən]	n. 过滤；浸透
dielectric [ˌdaɪɪ'lektrɪk]	n. 电介质；绝缘体
feedback ['fiːdbæk]	n. 反馈；成果，资料；回复
expenditure [ɪk'spendɪtʃə(r)]	n. 支出，花费；经费，消费额
radio frequency ['reidiəu 'friːkwənsi]	n. 射频

Ⅳ. **Phrases**

drip irrigation	滴灌
automated control system	自动控制系统
relay driver	继电器驱动器
solenoid control valves	电磁控制阀
remote computer	远程计算机
cellular network	蜂窝网

Exercises

Translate the following sentences into Chinese or English.

1. Static characteristics of sensor relate to issues such as how a sensor's output change in response to an input change, how selective the sensor is, how external or internal interferences can affect its response, and how stable the operation of a sensing system can be.

2. Sensor nodes usually consist of a processing unit with limited computational power and limited memory, sensors or MEMS (including specific conditioning circuitry), a communication device (usually radio transceivers or alternatively optical), and a power source usually in the form of a battery.

3. 无线传感网络的发展是出于战场监测等军事应用，当今这类网络也在许多行业得以应用，如工业过程监控、机器健康监控等。

4. 如果在传感器网络中使用集中结构并且中心节点出了故障，那么整个网络就会崩溃，但是可以使用分布式控制结构来增加这种传感器网络的可靠性。

Passage A 传感器特性及测量系统

Ⅰ.传感器的特性

传感器可以提供一个输出信号,以响应物理、化学以及生物的被测变量,通常还需要一个输入电信号进行操作。因此,与电子设备大致一样,它们的特性往往被写到传感器规格书(通常称为"规范")或者数据手册中。想要真正理解传感器,以及传感器测量同一个被测变量会有怎样的差异,需要理解传感器的运行特性。可惜的是,大多数传感器特性,还没有标准的定义。根据传感器领域,不同的传感器社区对于这些特性有不同的命名。制造商发布大量的不同的性能特性,进一步加剧了这种困惑。要识别与他们应用相关的某个性能,以及应用方法,对潜在客户来说愈加困难。下面的内容将以它们最常见的名称,描述这些特性,并提及这些代用名的出处。

1. 特性统计

(1) 灵敏度

灵敏度是指,单位量待测物质变化所致的响应量变化程度。如果传感器响应是线性的,那么灵敏度在整个传感器量程范围都会是恒定的,并且等于直线的斜率(如图4.1所示)。理想的传感器应该有持续恒定的灵敏度。如果传感器响应是非线性的,那么灵敏度会在传感器范围内发生变化,并且可以通过计算S关于x的导数求得,即dS/dx。

(2) 线性度

传感器的线性度是指,传感器的实际测量曲线偏离理想曲线的程度。图4.2显示了一种略微夸张的关系,关于理想的或最小二乘拟合线与实际测量或校准线之间的关系(注意在大多数情况下,静态曲线用作决定线性度,这有可能使动态线性度产生一些偏离)线性度通常是指非线性的百分比,表达式为:

$$\text{Nonlinearity}(\%) = \frac{D_{\text{in(max)}}}{\text{IN}_{f.s.}} \times 100$$

公式中非线性(%)指非线性的百分数,$D_{\text{in(MAX)}}$指最大输入偏移量,IN_{fs}指最大满量程输入。

方程式表达的静态非线性,经常受到环境因素的影响,包括温度、震动、噪声以及湿度。重要的是要知道,在怎样的条件下,其特性才有效,才能不受干扰,才使线性不会发生变化。

(3) 滞后现象

滞后现象不是指同一个被测变量的输出读数,而是依赖于传感器的轨道。

滞后可能会导致错误或不准确的读数。图4.3所示的是滞后系统中输出与输入的关系。从感测系统的显示可以看出,采用不同路径1或路径2,即使相同的输入,却得到两种不同的输出。

2. 动态特性

传感器动态特性可以告诉我们,如何很好地响应在输入中的变化。对于动态信号,传感器或测量系统必须能够快速响应,以跟上输入信号。

(1) 阶跃响应

响应速度:指示传感器(测量系统)如何快速地对输入变量的变化做出反应。(阶跃输入)上升时间:输入发生阶跃变化时,输出达到全响应的90%所耗费时间。时间常数:(一阶系统)输出达到其最大可能变化的63.2%的时间。调节时间:从采取输入步骤到特定频带的最后值时间内,可以完成输出。传递函数：一种描述传感器或系统性能的简单、简洁、完整的方法。公式$H(s) = Y(s)/X(s)$中$Y(s)$和$X(s)$分别是输出和输入的拉氏变换。有时传递函数以图形方式显示幅值和相位图与频率的关系（波特图）。

（2）频率响应

频率响应描述输出和输入变化与输入频率（正弦输入）的比例。

动态误差$\delta(\omega) = M(\omega) - 1$测量系统或传感器无法充分重建特定频率输入振幅。

带宽频带：$M(\omega) \geqslant 0.707(-3\text{dB})$

截止频率：系统响应已下降至稳定低频的0.707（-3dB）的频率。

Ⅱ. 测量系统

（1）初级感测元件

此原件首先接收被测介质的能量，再产生输出，该输出在某种程度上取决于被测量值。输出是一些物理变量，如位移或电压。仪器总是从被测介质中提取能量。被测量总是受到测量行为的干扰，这导致完美的测量在理论上是不可能的。设计较好的仪器可以减少这种负载效应。

（2）可调变换元件

保存原信号的信息内容时，可能需要将初级感测元件的输出信号转换为另一更合适的变量。此元件执行这样的功能。

（3）可变操纵元件

一种仪器可能需要某种物理上的信号来操纵某个信号。根据某些定义规则，我们进行操作，特别是操作数值上的变化，但保存的是物理性质的变量。此元件执行这样的功能。

（4）数据传输元件

实际上，仪器的功能元件在物理上是分离的，就这需要将数据从一个元件传输到另一个元件。此元件执行这样的功能。

（5）数据显示元件

如果将被测量的信息传达给某个人，用于监测、控制或分析目的，则必须形成一种可被人类感官识别的形式。此元件执行这种"传输"功能。

（6）数据存储元件

一些应用程序需要不同的数据存储，这样易于根据命令重新创建存储数据。

参考答案

1. 传感器的静态特性涉及一些问题，如传感器的输出变化如何响应其输入变化，传感器如何做选择，外部或内部干扰如何影响其响应，以及传感系统的操作如何稳定。

2. 传感器节点由以下部分组成：计算能力有限的处理器、有限的内存、传感器和MEMS（包括特定电路）、通信设备（通常是无线或光学收发器）以及电池形式的电源。

3. The development of wireless sensor networks was motivated by military applications such as battlefield surveillance; today such networks are used in many industrial and consumer application, such as industrial process monitoring and control, machine health monitoring, and so on.

4. If a centralized architecture is used in a sensor network and the central node fails, then the entire network will collapse. However the reliability of the sensor network can be increased by using distributed control architecture.

科技文翻译技巧二：增补与省略

增补与省略是翻译中最为常用的变通手段，增补也好，省略也好，都是增词不增义，减词不减义，反而能使意义更加明确，文字更加通达。

英译汉中的增补大致可以分为两种情形：一是根据原文上下文的意思、逻辑关系以及译文语言的行文习惯，在表达时增加原文字面上没有而意思上包含的字词；二是增补原文句法上的省略成分。

另外，有时候，英语中出现一些过于简洁的表意法，翻译时不做增补不足以说清楚意思，或者导致译文的生硬别扭，在这种情况下，译者也需酌情加以增补。例如：

(1) Rubber is found a good material for the insulation of cable.

译文：人们发现，橡胶是一种用于电缆绝缘的理想材料。

分析：英语有些被动句中没有行为主体，翻译时可增添适当的主语使译文通顺流畅。

(2) A router translates information from one network to another; it is similar to a super intelligent bridge.

译文：路由器把信息从一个网络传输给另一个网络，类似于超级智能的网桥。

Peter Newmark在《翻译教程》一书中，将"准确"（accuracy）和"简练"（economy）视为翻译的基本要求；加上汉语又是意合语言，尤其讲究简洁顺畅，因此我们做英译汉时，还要学会掌握省略法，也就是说，要学会舍去原文中需要而译文中不需要的成分。例如：

(3) The electric power industry is primarily concerned with energy conversion and distribution.

译文：电力工业同能源的转化和分配密切相关。

分析：英语中的系动词be, become等可根据具体情况在翻译中省译。

(4) A bridge is a device that allows you to segment a large network into two smaller, more efficient networks.

译文：网桥是一个设备，可以把一个大网段分割成两个更小、更有效的网段。

Unit 5

Wi-Fi

Passage A   An Overview of Wi-Fi

Passage B   CSMA/CA Mechanism

Passage C   Surveillance System Applications Over Public Transportation

Passage A An Overview of Wi-Fi

Ⅰ. An Introduction

Wi-Fi or WiFi is a technology that allows electronic devices to connect to a wireless LAN (WLAN) network, mainly using the 2.4 gigahertz (12 cm) UHF and 5 gigahertz (6 cm) SHF ISM radio bands. Access to a WLAN is usually password protected, but may be open, which allows any device within its range to access the resources of the WLAN network.

The Wi-Fi Alliance defines Wi-Fi as any "wireless local area network" (WLAN) product based on the Institute of Electrical and Electronics Engineers' (IEEE) 802.11 standards. However, the term "Wi-Fi" is used in general English as a synonym for "WLAN" since most modern WLANs are based on these standards. "Wi-Fi" is a trademark of the Wi-Fi Alliance. The "Wi-Fi Certified" trademark can only be used by Wi-Fi products that successfully complete Wi-Fi Alliance interoperability certification testing.

Devices which can use Wi-Fi technology include personal computers, video-game consoles, smartphones, digital cameras, tablet computers and digital audio players. Wi-Fi compatible devices can connect to the Internet via a WLAN network and a wireless access point. Such an access point (or hotspot) has a range of about 20 meters (66 feet) indoors and a greater range outdoors. Hotspot coverage can be as small as a single room with walls that block radio waves, or as large as many square kilometres achieved by using multiple overlapping access points.

Ⅱ. Wi-Fi Working Principle

1. Wi-Fi Working Concepts

(1) Radio Signals

Radio Signals (as shown in Figure 5.1) are the keys, which make Wi-Fi networking possible. These radio signals transmitted from Wi-Fi antennas are picked up by Wi-Fi receivers, such as computers and cell phones that are equipped with Wi-Fi cards. Whenever, a computer receives any of the signals within the range of a Wi-Fi network, which is usually 300 ~ 500 feet for antennas, the Wi-Fi card reads the signals and thus creates an internet connection between the user and the network without the use of a cord.

Access points, consisting of antennas and routers, are the main source that transmit and receive radio waves. Antennas work stronger and have a longer radio transmission with a radius of 300~500 feet, which are used in public areas while the weaker yet effective router is more suitable for homes with a radio transmission of 100~150 feet.

(2) Wi-Fi Cards

You can think of Wi-Fi cards (as shown in Figure 5.2) as being invisible cords that connect your computer to the antenna for a direct connection to the internet.

(3) Wi-Fi Hotspots

A Wi-Fi hotspot is created by installing an access point to an internet connection. The access point transmits a wireless signal over a short distance. It typically covers around 300 feet. When a Wi-Fi enables device such as a Pocket PC encounters a hotspot, the device can then connect to that network wirelessly.

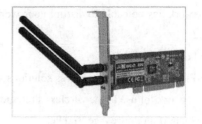

Figure 5.1　Radio Signals　　　　　　　　Figure 5.2　Wi-Fi Card

Most hotspots are located in places that are readily accessible to the public such as airports, coffee shops, hotels, book stores, and campus environments. 802.11b is the most common specification for hotspots worldwide. The 802.11g standard is backwards compatible with11b but 11a uses a different frequency range and requires separate hardware such as an a, a/g, or a/b/g adapter. The largest public Wi-Fi networks are provided by private internet service providers (ISPs); they charge a fee to the users who want to access the internet.

2. Working Principle

Basically, the working principle of Wi-Fi can be described as a computer's or other compatible device's wireless adapter translates data into a radio signal (2.4 and 5 GHz radio bands) and transmits it to a wireless router by using an antenna. When the wireless router receives the signal, it will decode it and send the data to the Internet by using a physical, wired Ethernet connection (as shown in figure 5.3). The working process also works in reverse. When the router receives data from the Internet, it will translate the data into a radio signal and send it to the computer's or other compatible device's wireless adapter.

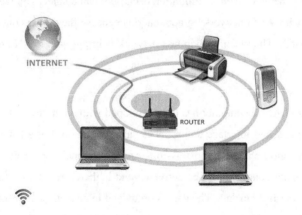

Figure 5.3　Working Principle of Wi-Fi

3. Features of Wi-Fi Technology

(1) Unmatched Mobility and Elasticity

Wi-Fi is allowing new intensity of connectivity without giving up functions. Wi-Fi introduced various types of utilities such music streamers that transmit your music to speakers without any wire you can also play music from the remote computer or any other attached to the network. The most important now you can play online radio. Wi-Fi technology system is rather remarkable, you can download songs, send emails and transfer files expediently at skyscraping speed and you can move your computer easily because your

Wi-Fi network has no cable to disrupt your work so we can say that it is quite easy, helpful and most of all expedient.

(2) Fortress Technology

Wi-Fi providing secure wireless solutions support the growth and release of a prototype mobile ad hoc wireless network for use in the wireless strategic skirmish.

(3) Support an Entire Age Bracket

Wi-Fi technology has several advantages and it support an entire age bracket and create a connection between components on the same network and has ability to transfer data between the devices and enables different kinds of devices such as game, MP3 player, PDA's and much more!

(4) Convenient and Everywhere

Wi-Fi is a convenient technology and where the range station exists you are online during travel you can equip with a Wi-Fi network and set up shop anyplace. You will automatically connect with internet if you are near hotspot. These days Wi-Fi exist everywhere with all its wonders.

(5) Faster and More Secure

With Wi-Fi you can get high speed of internet because it is very fast than DSL and Cable connection you can establish a Wi-Fi network in small space now you don't need any professional installation just connect to a power outlet with an Ethernet cord, and start browsing. Wi-Fi security system for Threats makes it more renewable and its tool protect your VPN and secure web page. You can easily configure the device to take better performance. The standard devices, embedded systems and network security make it more powerful.

(6) Wi-Fi with No Limitation

You can use a "Wi-Fi" network with no limitation because it can connect you worldwide. You can easily reach your requirements with Wi-Fi networking applications because the power consumption is very high as compared to other bandwidth. The vision of wireless network is bright with Pre-N products and high quality media streaming.

III. Wi-Fi Application

To connect to a Wi-Fi LAN, a computer has to be equipped with a wireless network interface controller. The combination of computer and interface controller is called a station. For all stations that share a single radio frequency communication channel, transmissions on this channel are received by all stations within range. The transmission is not guaranteed to be delivered and is therefore a best-effort delivery mechanism. A carrier wave is used to transmit the data. The data is organized in packets on an Ethernet link, referred to as "Ethernet frames". The Wi-Fi product contains a router with an 802.11 b/g radio and two antennas is shown as Figure 5.4.

Figure 5.4　The Wi-Fi praduct contains a router with an 802.11b/g radio and two antennas

Wi-Fi technology may be used to provide Internet access to devices that are within the range of a wireless network that is connected to the Internet. The coverage of one or more interconnected access points (hotspots) can extend from an area as small as a few rooms to as large as many square kilometres. Coverage in the larger area may require a group of access points with overlapping coverage. For example, public outdoor Wi-Fi technology has been used successfully in wireless mesh networks in London, UK. An international example is FON.

Wi-Fi provides service in private homes, businesses, as well as in public spaces at Wi-Fi hotspots set up either free-of-charge or commercially, often using a captive portal webpage for access. Organizations and businesses, such as airports, hotels, and restaurants, often provide free-use hotspots to attract customers.

Routers that incorporate a digital subscriber line modem or a cable modem and a Wi-Fi access point, often set up in homes and other buildings, provide Internet access and internetworking to all devices connected to them, wirelessly or via cable.

Similarly, battery-powered routers may include a cellular Internet radiomodem and Wi-Fi access point. When subscribed to a cellular data carrier, they allow nearby Wi-Fi stations to access the Internet over 2G, 3G, or 4G networks using the tethering technique. Many smartphones have a built-in capability of this sort, including those based on Android, BlackBerry, Bada, iOS (iPhone), Windows Phone and Symbian, though carriers often disable the feature, or charge a separate fee to enable it, especially for customers with unlimited data plans. "Internet packs" provide standalone facilities of this type as well, without use of a smartphone; examples include the MiFi and WiBro-branded devices. Some laptops that have a cellular modem card can also act as mobile Internet Wi-Fi access points.

1. City-wide Wi-Fi

In the early 2000s, many cities around the world announced plans to construct city-wide Wi-Fi networks. There are many successful examples; in 2004, Mysore became India's first Wi-Fi-enabled city. A company called Wi-FiyNet has set up hotspots in Mysore, covering the complete city and a few nearby villages.

In 2005, St. Cloud, Florida and Sunnyvale, California, became the first cities in the United States to offer city-wide free Wi-Fi (from MetroFi). Minneapolis has generated $1.2 million in profit annually for its provider.

In May 2010, London, UK, Mayor Boris Johnson pledged to have London-wide Wi-Fi by 2012. Several boroughs including Westminster and Islington already had extensive outdoor Wi-Fi coverage at that point.

Officials in Republic of Korea's capital are moving to provide free Internet access at more than 10,000 locations around the city, including outdoor public spaces, major streets and densely populated residential areas. Seoul will grant leases to KT, LG Telecom and SK Telecom. The companies will invest $44 million in the project, which was to be completed in 2015. An outdoor Wi-Fi access point is shown as Figure 5.5.

2. Campus-wide Wi-Fi

Many traditional university campuses in the developed world provide at least partial Wi-Fi coverage. Carnegie Mellon University built the first campus-wide wireless Internet network, called Wireless Andrew, at its Pittsburgh campus in 1993 before Wi-Fi branding originated By February 1997 the CMU zone was fully operational. Many universities collaborate in providing Wi-Fi access to students and staff through the eduroam international authentication infrastructure.

Figure 5.5 An outdoor Wi-Fi access point

3. Direct computer-to-computer communications

Wi-Fi also allows communications directly from one computer to another without an access point intermediary. This is called ad hoc Wi-Fi transmission. This wireless ad hoc network mode has proven popular with multiplayer handheld game consoles, such as the Nintendo DS, PlayStation Portable, digital cameras, and other consumer electronics devices. Some devices can also share their Internet connection using ad hoc, becoming hotspots or "virtual routers".

Similarly, the Wi-Fi Alliance promotes the specification Wi-Fi Direct for file transfers and media sharing through a new discovery- and security-methodology. Wi-Fi Direct launched in October 2010.

Another mode of direct communication over Wi-Fi is Tunneled Direct Link Setup (TDLS), which enables two devices on the same Wi-Fi network to communicate directly, instead of via the access point.

Ⅳ. New Words

password ['pɑːswɜːd]	n.	密码；口令
radio ['reɪdɪəʊ]	n.	收音机；无线电广播设备
modem ['məʊdem]	n.	调制解调器
certification [ˌsɜːtɪfɪ'keɪʃən]	n.	证明，保证；检定
digital ['dɪdʒɪt(ə)l]	adj.	数字的；手指的
router ['raʊtə]	n.	路由器
radius ['reɪdɪəs]	n.	半径，半径范围；辐射光线；有效航程

Ⅴ. Phrases

electronic devices	电子设备
radio signals	无线电信号
Fortress Technology	堡垒技术

Ⅵ. Abbreviations

WLAN	Wireless Local Area Network	无线局域网
UHF	Ultra High Frequency	超高频
DSL	Digital Subscriber Loop	数字用户环路
VPN	Virtual Private Network	虚拟专用网络

Passage B CSMA/CA Mechanism

Ⅰ. CSMA/CA mechanism

Collision avoidance is used to improve the performance of the CSMA method by attempting to divide the channel somewhat equally among all transmitting nodes within the collision domain.

1. Carrier Sense

Prior to transmitting, a node first listens to the shared medium (such as listening for wireless signals in a wireless network) to determine whether another node is transmitting or not. Note that the hidden node problem means another node may be transmitting which goes undetected at this stage.

2. Collision Avoidance

If another node was heard, we wait for a period of time for the node to stop transmitting before listening again for a free communications channel.

3. Request to Send/Clear to Send (RTS/CTS)

It may optionally be used at this point to mediate access to the shared medium. This goes some way to alleviating the problem of hidden nodes because, for instance, in a wireless network, the Access Point only issues a *Clear to Send* to one node at a time. However, wireless 802.11 implementations do not typically implement RTS/CTS for all transmissions; they may turn it off completely, or at least not use it for small packets (the overhead of RTS, CTS and transmission is too great for small data transfers).

4. Transmission

If the medium was identified as being clear *or* the node received CTS to explicitly indicate it can send, it sends the frame in its entirety. Unlike CSMA/CD, it is very challenging for a wireless node to listen at the same time as it transmits (its transmission will dwarf any attempt to listen). Continuing the wireless example, the node awaits receipt of an acknowledgement packet from the Access Point to indicate the packet was received and check summed correctly. If such acknowledgement does not arrive in a timely manner, it assumes the packet collided with some other transmission, causing the node to enter a period of binary exponential backoff prior to attempting to re-transmit.

Although CSMA/CA has been used in a variety of wired communication systems, it is particularly beneficial in a wireless LAN due to a common problem of multiple stations being able to see the Access Point, but not each other. This is due to differences in transmit power, and receive sensitivity, as well as distance, and location with respect to the AP. This will cause a station to not be able to "hear" another station's broadcast. This is the so-called "hidden node", or "hidden station" problem. Devices utilizing 802.11 based standards can enjoy the benefits of collision avoidance (RTS / CTS handshake, also Point coordination function), although they do not do so by default. By default they use a Carrier sensing mechanism called "exponential backoff", or (Distributed coordination function) that relies upon a station attempting to "listen" for another station's broadcast before sending. CA, or PCF relies upon the AP (or the "receiver" for Ad hoc networks) granting a station the exclusive right to transmit for a given period of time after requesting it (Request to Send / Clear to Send).

Ⅱ. RTS/CTS mechanism in IEEE 802.11

In this section, as a typical example of the implementation of the RTS/CTS mechanism, we briefly describe the DCF mode of the IEEE 802.11 Wireless LAN protocol.

In the DCF mode, a node may transmit a packet using one of the following two methods: the basic access method or the RTS/CTS method. In the basic access method, a node transmits a DATA packet if it senses the channel to be idle. The receiver, upon receiving an error-free packet, returns an ACK. If the transmitting node does not get an ACK back, it enters into back-off and retransmits after the back-off period.

The basic access method suffers from the well-known hidden node problem. In order to address the issue, IEEE 802.11 supports the RTS/CTS access control mode. The RTS/CTS access mode is a combination of carrier sensing. Figure 5.6 illustrates the scheme. When a node A wants to send a packet to node B, it initially sends a small packet called Request-to-Send (RTS). Upon correctly receiving the RTS, node B responds with another small packet called Clear-to-Send (CTS). After receiving the CTS, node A sends the DATA packet to node B. If node B receives the DATA packet correctly, it sends an Acknowledgment (ACK) back to node A. Any node that hears an RTS or CTS is prohibited from transmitting any signal for a period that is encoded in the duration field of the received RTS or CTS. The duration fields in RTS and CTS are set such that nodes A and B will be able to complete their communication within the prohibited period. The deferral periods are managed by a data structure called the Network Allocation Vector (NAV). Finally, if a node does not get a response to an RTS or a DATA packet, it enters into an exponential backoff mode.

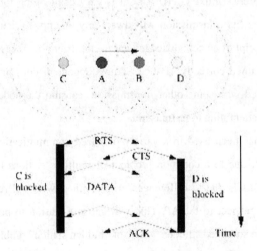

Figure 5.6　IEEE 802.11 MAC. The lower half depicts the time-line.
The dark bars below node C and D indicates their NAV.

Ⅲ. New Words

Transmission [træns'mɪʃən]	n.	传动装置，变速器；传递；传送；播送
Collision [kə'lɪʒ(ə)n]	n.	碰撞；冲突
Carrier ['kærɪə]	n.	载体；运送者
Node [nəʊd]	n.	节点
Mediate ['miːdɪeɪt]	vi.	调解；斡旋

Ⅳ. Phrases

Carrier Sense	载波侦听
Collision Avoidance	冲突避免

Ⅴ. Abbreviations

CSMA	Carrier Sense Multiple Access	载波侦听多路访问
IEEE	Institute of Electrical and Electronic Engineers	电气与电子工程师协会

Passage C Surveillance System Applications Over Public Transportation

Ⅰ. Surveillance system architecture

In this section we present the monitoring device that is being deployed in our implementation, providing an overview analysis of its basic modules. Its architecture must be highly modular, it should be possible to substitute modules easily so as to adapt into various operative environmental challenges such as a tunnel. The main core is a three layer stackable PC/104+board. Each stack must contain at least one mother board or CPU, which acts as a controller for the rest peripheral components. The generic monitoring and comprises the following components: a stackable PC/104+ board, an explosion sensor, a frame grabber/encoder and the Wi-Fi - WiMAX interface module.

The most important attributes provided for the PC/104+board are the Processing speed (equipped with a 1.4 GHz processor), Storage capability (supports up to 8GB Removable Industrial ATA/IDE Flash Disk (4GB Standard)), Power Consumption (power consumption is strongly related to the processor speed range, from 10 to 20 W) and Operating Temperature Range (from -40 °C to $+85$ °C). In the case that an analogue output camera is selected, an extra frame grabber will be needed in order to extract and digitize video frames from analogue signal.

The triggering module may comprise two different types of sensor connected to the CPU board. The CPU board provides the control integration functions. It constantly checks for a pressure increase and/or very sudden start or stop that is beyond a user-selectable threshold value, determines if it fits the parameters associated with an explosion and sends a signal (trigger) to the system controller.

Alarm triggering mode under an emergency situation is triggered by an explosion, the triggering is immediately propagated to the devices 01 and 02 via IEEE 802.11b/g/e interface and, at the same time, an alarm is sent to the central station through IEEE 802.16e mobile WiMAX interface. The cyclic buffer of all three devices is frozen so that the last XX pictures are kept in memory. The devices that have successfully uploaded their data become available for on-demand real-time video link between the central station and the affected space. Thus each monitoring device propagates the triggering to its neighbouring devices (i.e. the devices monitoring the rest of the wagons of the train) so that the system can capture and make available to the authorities (rescuers, police etc.) as complete set of information as possible. Once triggered, each device needs to communicate with the neighbouring devices in order to forward the uploaded data (frames - real time video) therefore, each wagon of the train is equipped with a Wi-Fi IEEE 802.11b/g/e USB dongle responsible for implementing the intra wireless network scenario.

In order to be transmitted across the tunnel, an inter mobile WiMAX (IEEE 802.16e)network is implemented the edges of the train providing connectivity with the WiMAX Base Stations located at the two edges of the tunnel respectively. The utilized IEEE 802.16e protocol is the most appropriate protocol on addressing harsh propagation environments, where NLOS (Non-Line-of-Sight) appearance could cause severe interference into the propagation signal, providing reliability, and inter operability, attributes necessary for a robust surveillance network.

II. Communication module

A tunnel environment appears to be an extremely demanding area in terms of wireless communications. Therefore, the communication of the implemented surveillance system must ensure smooth cooperation between different wireless protocols providing continuously connectivity across the tunnel. Mobility of railways inside the tunnel structure adds its own strict parameters that are needed to be taken into account in order to establish the preferred wireless network communication between the mobile metro coaches and each station. Harsh propagation conditions during the loss of LOS (Line-of-Sight) occasionally and especially in terms of an emergency situation (explosion event) can affect severely the network connection. Inter symbol interference and possible occurrence of blind spots may also lead to severe derogation performance. In order to overcome this impediment it is necessary to have a combination of wireless protocols standards such as IEEE 802.11b/g/e and mobile WiMAX IEEE802.16e to overcome multipath and NLOS (Non-Line-of-Sight) consequences. The view of Intra and Inter wireless network over the coaches is presented in Figure. 5.7. There exist two distinct basic communication functionalities.

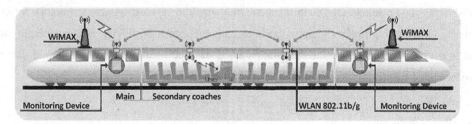

Figure 5.7 Train Surveillance wireless network

III. Communication modules and functionality

The basic equipment introduced in our experiment is a Subscriber unit (SU) with dual WLAN and WiMAX interfaces located among the two edges of the train. An uplink is created between 4 I.EEE. 802.11b/g/e USB dongles the SU's(AP) Access Point (WLAN SU interface). The received WLAN data packets are retransformed to WiMAX data packets and transmitted to the WiMAX Base station. The proposed WiMAX interface of the Subscriber unit supports the unlicensed band of 5.47–5.95 GHz (46 channels) and offers point-to-multipoint wireless communication delivering data and video during in-progress events with fast handoff at speeds of 120 km/h. It can also provide Dynamic Data Rate Selection and automatically compensates for temporary link degradation while maintaining robust connectivity, allowing bandwidth intensive applications, such as high-definition video streaming, in mobile environments. The IEEE 802.11b/g/e USB dongles operating at 2.4 GH zare equipped with an external 5dBi Omni-directional antenna. Due to the fact that our experiment includes H.264 video data streaming applications, the Wi-Fi USB dongles must support the IEEE 802.11e standard providing a set of Quality of Service enhancements for wireless LAN applications. The standard is considered of critical importance for delay sensitive applications, such as Voice over Wireless LAN and streaming multimedia.

IV. Intra domain communication

The proposed IEEE 802.11b/g is the most appropriate protocol for utilizing the intra communication scenario, due to its robustness and credibility in intermediate distances, providing connectivity range among the train coaches. The train consists of 4 coaches each one is equipped with a Wi-Fi USB dongle, while the first and the last coach apart from a Wi-Fi USB dongle are also equipped with an SU with AP WLAN interface. The intermediate coaches are only equipped with 2 Wi-Fi USB dongles located inside the coach within a 20 m distance between them. Therefore, the Intra communication scenario is implemented via IEEE 802.11b/g/e Wi-Fi USB dongles that can upload multimedia data, captured by the IP's cameras of each coach, to both SUs - APs (WLAN interface) that act as a central transmitter and receiver for the intra network located at the two edges of the train respectively.

Ⅴ. Inter domain communication

Analyzing the Inter communication module we introduce the mobile WiMAX backhaul protocol mainly responsible for transporting data/voice traffic from a WiMAX SU to a WiMAX BS. For this reason in order to develop a mobile WiMAX network the WiMAX IEEE 802.16e interface of both SUs, located across the train edges, implements a WiMAX backhaul network with the BS1 and BS2 BaseStations respectively that are located towards the tunnel entrances. IEEE 802.16e capability to provide very high capacity links on both uplink (UL) and downlink (DL) confirms the reason why it is the most suitable technology for in-tunnel applications. Additionally, it offers the possibility to communicate with moving nodes at ranges sufficient so as not to have repeaters that hop the information along the metro galleries. The next hop of data is to be uploaded through the Base Stations via WiMAX to a central station.

Ⅵ. New Words

surveillance [sɜːˈveləns]   n. 监督；监视
architecture [ˈɑːkɪtektʃə]   n. 架构
monitoring [ˈmɒnɪtərɪŋ]   n. 监视，监控；检验，检查
modules [ˈmɔdʒulz]   n. 模块
stackable [ˈstækəbl]   adj. 易叠起堆放的；可叠起堆放的
analogue [ˈænəlɒg]   n. 模拟
parameters [pəˈræmɪtəz]   n. 参数，参量；界限；因素，特征

Ⅶ. Abbreviations

LOS Line-of-Sight   视线

Exercises

Translate the following sentences into Chinese or English.

1. Wi-Fi is a popular technology that allows an electronic device to exchange data wirelessly (using radio waves) over a computer network.

2. Routers that incorporate a digital subscriber line modem or a cable modem and a Wi-Fi access point often set up in homes and other buildings, provide Internet access and internetworking to all devices connected to them, wirelessly or via cable.

3. 计算机必须配置无线网络接口控制器，才能连接到Wi-Fi网络。

4. Wi-Fi可以在私人住宅、繁华大街、独立商业机构和公共空间提供服务。

参考译文 Passage A Wi-Fi技术概述

Ⅰ. 概述

Wi-Fi（或WiFi）是一种允许电子设备连接到无线局域网（WLAN）的技术，通常使用2.4G UHF或5G SHF ISM射频频段。无线局域网通常是有密码保护的；但也可是开放的，开放的无线局域网就允许任何在无线局域网范围内的设备连接。

无线联盟把Wi-Fi定义为，基于电气和电子工程师协会（IEEE）802.11标准的"无线局域网"产品。然而，Wi-Fi一词在普通英语中是"WLAN"的近义词，因为大多数现代无线局域网都基于这些标准。Wi-Fi是无线联盟的商标，其商标"Wi-Fi Certified"只用于成功通过无线联盟互操作性认证测试的无线产品。

可以使用Wi-Fi的设备包括个人计算机、游戏机、智能手机、数码相机、平板电脑以及数字音频播放器。Wi-Fi兼容设备可以通过无线局域网和无线接入点连接互联网。一个接入点（或者热点）的室内范围是20m（66英尺），室外的范围更大。热点的覆盖范围可以小到一间房间的大小，墙壁会阻止无线电波，也可以大到几平方公里范围内，实现多个重叠接入点。

Ⅱ. Wi-Fi工作原理

1. Wi-Fi工作概念

（1）无线电信号

无线电信号（见图5.1）是Wi-Fi网络能否可用的关键。无线电信号通过Wi-Fi接收器的Wi-Fi天线进行传输，例如，计算机和手机都配有Wi-Fi卡。Wi-Fi网络范围内，计算机任何时候都能接收到信号，一般在天线附近的300英尺到500英尺，Wi-Fi卡会读取信号，然后创建互联网连接用户和网络，不需要任何网线。

接入点是传输和接收无线电波的主要来源，由天线和路由器组成。天线越好，无线电传播的范围就越广，可达到300～500英尺，这种适用于公共场所，较弱一点的路由器更适合家用，其无线电传播范围是100～150英尺。

（2）Wi-Fi卡

你可以想象，Wi-Fi卡就像一根隐形的绳子，通过天线直接将计算机与互联网相连接，如图5.2所示。

（3）Wi-Fi热点

Wi-Fi热点是通过安装接入点创建互联网连接，接入点在短距离内传输无线信号，通常范围在300英尺之内。启用Wi-Fi时，设备就能连接到热点，也就能连接到无线网络，例如掌上电脑。大多数热点都在易于公众使用的地方，例如机场、咖啡馆、宾馆、书店和校园。802.11b是全球热点最常见的规格。802.11g标准可以兼容11b，但是与11a工作于不同频段，并且需要单独的硬件，例如，a、a/g、a/b/g适配器。最大的公共Wi-Fi网络由私人互联网供应商提供，并向使用者收取一定费用。

2. 工作原理

Wi-Fi的工作原理大致可以描述为，计算机或其他兼容设备的无线适配器把数据转化为无线信号（2.4 and 5GHz频段），并通过天线把无线信号传送到无线路由器。路由器接收到信号后，进行解码，然后通过连接以太网的形式把数据传送给互联网（见图5.3）。其工作过程也同样如此，路由器从互联网接受数据，然后将数据转化成无线信号，再发送给计算机或其他兼容设备的无线适配器。

3. 无线技术的特征

（1）无与伦比的移动性和灵活性

Wi-Fi可在不影响现运行的某种功能用途下，进行其他功能用途连接。Wi-Fi推出各类程序，例如音乐数据流，它可以不通过任何电线传输音乐，你也可以通过计算机远程或其他网络连接播放音乐。最重要的是还可以使用在线电台。Wi-Fi技术系统很卓越，下载音乐、发送邮件、传输文件都很方便快速，你也可以轻松地移动计算机，因为Wi-Fi网络不像电缆那样会干扰工作，所以我们可以说这是方便有用、最有利的选择。

（2）堡垒技术

Wi-Fi提供安全的无线解决方案，支持原型手机的增长和释放，专设的无线网可用于解决无线战略冲突。

（3）支持所有用户群体

Wi-Fi技术有多个优点，它支持所有用户群体，并可以创建相同网络中的组件连接，还可以进行设备与设备之间的数据传输，例如，游戏、MP3播放器、掌上电脑等。

（4）方便和无处不在

Wi-Fi是种方便的技术，如果在有信号站存在的地方，带上Wi-Fi网络设备，整个旅程你都可以在线上，并且在任何地方都能购物。只要附近有热点，你就可以自动连接网络，有Wi-Fi就有奇迹。

（5）快速、安全

使用Wi-Fi可以提高网络速度，因为Wi-Fi比DSL和电缆连接都快，你可以在小空间内建立Wi-Fi网络，并且不需要任何专业的装置，只需要将以太网电缆连接到电源，就可以开始浏览网页了。Wi-Fi安全系统的威胁制造出更多再生资源，其工具可以保护虚拟私人网络及网页安全。你可以轻松安装设备获取更好的性能。标准设备、嵌入式系统和网络安全使它更强大。

（6）Wi-Fi没有限制

使用Wi-Fi没有任何限制，因为你可以在任何地方连接到Wi-Fi。在Wi-Fi网络下，你可以轻松使用设备上的任何应用，因为其功率消耗要高于其他的频带宽度。拥有Pre-N产品和高质量媒体数据流，无线网络的前景是美好的。

III. 无线网络应用

要连接到Wi-Fi网络，计算机必须配备有无线网络接口控制器。计算机和接口控制器的组合被称为站。对于共享一个射频通信频道的所有站，该通道上的任何传输是由范围内的所有站接收。发送不保证信息能够被传递，因此是一种最大努力的递送机制。载波被用来传输数据。数据以分组组织方式在以太网链路上，被称为"以太网帧"。

Wi-Fi技术可用于在无线网络范围内连接到因特网的装置提供互联网接入。一个或多个互连的接入点（热点）的覆盖范围可以从小如几个房间一样大，到许多平方公里的延伸。大面积覆盖，可能需要许多组接入点的重叠覆盖。例如，在英国伦敦，公共室外Wi-Fi技术已经成功应用在无线网状网络中。纤维光学网络便是一个国际性的例子。

无线网络为私人住宅，企业，以及设立的免费或商用Wi-Fi热点的公共场所提供服务，通常只需访问门户网页便可入网。组织和企业，如机场，酒店和餐馆，往往提供免费使用的热点，以吸引顾客。

路由器集成了数字用户线调制解调器或电缆调制解调器和Wi-Fi接入点，通常安装在家庭和其他建筑物内。通过无线或电缆，为所有设备提供互联网接入和互联网络连接。

同样的，电池供电的路由器可以包括蜂窝网络无线调制解调器和Wi-Fi接入点。当订阅了蜂窝数据载体，它们允许附近的Wi-Fi站使用圈养技术，通过2G，3G或4G网络接入互联网。许多智能手机都有一个对客户提供无限数据计划的内置功能，尽管运营商通常禁用该功能，或收取不同的费用，以启用它。其包括基于Android、BlackBerry、Bada、iOS设备（iPhone）、Windows手机和Symbian。没有使用智能手机的情况下，"互联网包"提供这种类型的独立实施，其实例包括MiFi和WiBro品牌的设备。有蜂窝调制解调器卡的一些笔记本电脑也可以作为移动互联网的Wi-Fi接入点。

1. 城市Wi-Fi覆盖

21世纪初期，世界各地的许多城市都宣布计划构建全市Wi-Fi网络。图5.5显示了一个户外的Wi-Fi接入点有很多成功的例子；在2004年，迈索尔成为印度第一个支持Wi-Fi的城市。一家名为Wi-FiyNet的公司，在迈索尔已成立了热点，覆盖整个城市和一些附近的村庄。

2005年，圣克劳德市，佛罗里达州和加利福尼亚州桑尼维尔市，成为美国第一个提供全市范围内免费的Wi-Fi的城市（从MetroFi）。明尼阿波利斯为它的供应商每年产生的利润为一百二十万美元。

2010年5月，伦敦市长鲍里斯·约翰孙承诺到2012年伦敦Wi-Fi普及，几个自治市镇包括威斯敏斯特和伊斯灵顿已经有了广泛的Wi-Fi覆盖。

在韩国首都的官员正在试图向城市周围超过10,000个地点提供免费上网，包括户外公共场所、主要街道和人口密集的居住区。首尔将向KT，LG电信和SK电信等授予租约。这些公司将为此项目投资4400万美元，此项目预计在2015年完成。户外Wi-Fi接入点如图5.5所示。

2. 校园的Wi-Fi覆盖

在发达国家，许多传统的大学校园至少提供部分Wi-Fi覆盖。卡内基·梅隆大学建立了第一个校园范围内的无线互联网络，称为无线安德鲁。这是在其匹兹堡校区于1993年，在Wi-Fi品牌之前提出的，于1997年2月CMU区全面运作。许多大学在通过eduroam国际认证基础设施，向学生和教师人员提供Wi-Fi接入。

3. 计算机对计算机直接通信

Wi-Fi无线通信也可以不需通过接入点，直接连接两台计算机。这就是所谓自组织模式的Wi-Fi传输。这种无线自组织网络模式受到掌上游戏机、数码相机和其他消费性电子设备的欢迎。一些设备也可以使用自组织，成为热点或"虚拟路由器"分享他们的互联网连接。

同样，Wi-Fi联盟推动一个新的安全方法规范，称为Wi-Fi Direct，直接进行文件传输和媒体共享。Wi-Fi Direct于2010年10月推出。

通过Wi-Fi直接沟通的另一种模式是隧道直接链路建立（TDLS），从而允许同一Wi-Fi网络上的两个设备直接通信，而不通过接入点。

参考答案

1. Wi-Fi 是一种流行技术，允许电子设备（使用无线电波）通过计算机网络进行无线交换数据。

2. 带有数字用户线路调制解调器或电缆调制解调器的路由器与 Wi-Fi 访问接入点通常设置在房间里或大楼内，为有线或无线方式连接到的所有设备，提供网络连接和因特网访问服务。

3. To connect to a Wi-Fi LAN, a computer has to be equipped with a wireless network interface controller.

4. Wi-Fi provides service in private homes, high street chains and independent businesses, as well as in public spaces.

科技文翻译技巧三：转性与变态

转换词性是翻译中最为常用的一种变通手段，是突破原文词法、句法格局，化阻滞为通达的重要方法。离开必要的词性转换，势必会导致译文生硬拗口，甚至晦涩难懂。当然，词性转换要本着一个原则，即不违背原文的意思，有助于译文的通顺流畅。

从理论上来说，翻译中的词性转换是没有限制的，比如说，名词可以转换成动词、形容词、副词等，动词可以转换成名词、形容词、副词等，而形容词、副词也可以转换成名词、动词等，不一而足。不过，从实践来看，英汉语的词性转换也有一定的规律，最明显的一点，就是英语比较喜欢多用名词和介词，而汉语则是动词用得多一些。因此，我们在做英译汉时，词性的转换比较多地表现在名词、介词变动词上。

1. 介词转换成动词

(1) By radar people can see things beyond the visibility of them.

译文：利用雷达，人们能看见视线外的物体。

分析：By radar 译成"利用雷达"。

(2) Power is needed to stall the armature against inertia.

译文：为使电枢克服惯性而制动，需要一定的能量。

分析：against 译成"克服"。

2. 名词转换成动词

(1) In the dynamo, mechanical energy is used for rotating the armature in the field.

译文：在直流发电机中机械能用来使电枢在磁场中转动。

分析：rotating 译成"使……转动"。

(2) The flow of electrons is from the negative to the positive.

译文：电子是从负极流向正极。

分析：把"flow"由名词转译为动词"流向"，全句更简洁。

在科技英语中被动语态广泛使用，以突出强调所要论证和说明的科技问题。把英语的被动语态译成汉语时，一方面为了保留原文的风格特点，有些句子仍要译成被动语句。主要使用的手段是在位于行为主体前加"被""由""受""为……所"和"是……的"等字。另一方面，为了使句子通顺、明确，要把英语的被动语态译成汉语的主动语态。主要方法有更换主语、增译主语或译成汉语的无主句。因此翻译时，要根据需要采取灵活多样的译法。

1. 译成被动语态

Atoms were considered to be indivisible units of matter.

译文：原子曾被认为是物质不可分的最小单位。

分析：加"被"字表达被动意义。

2. 译成主动语态

More than one hundred elements have been found by chemical workers at present.

译文：目前，化学工作者已经发现了一百多种元素。

分析：更换主语，把被动语态译为主动语态。

Unit 6

Wireless Sensor Networks

Passage A   ZigBee Technology

Passage B   IEEE 802.15.4 Technology and 6LoWPAN

Passage C   Online Monitoring System for Transmission Lines and Sensing of Traffic Flows

Passage A ZigBee Technology

I. Overview

ZigBee is a low-cost, low-power, wireless mesh network standard targeted at the wide development of long battery life devices in wireless control and monitoring applications. Zigbee devices have low latency, which further reduces average current. ZigBee chips are typically integrated with radios and with microcontrollers that have between 60-256 KB of flash memory. ZigBee operates in the industrial, scientific and medical (ISM) radio bands: 2.4 GHz in most jurisdictions worldwide; 784 MHz in China, 868 MHz in Europe and 915 MHz in the USA and Australia. Data rates vary from 20 kbit/s (868 MHz band) to 250 kbit/s (2.4 GHz band).

The ZigBee network layer natively supports both star and tree networks, and generic mesh networking. Every network must have one coordinator device, tasked with its creation, the control of its parameters and basic maintenance. Within star networks, the coordinator must be the central node. Both trees and meshes allow the use of ZigBee routers to extend communication at the network level.

ZigBee is one of the global standards of communication protocol formulated by the significant task force under the IEEE 802.15 working group. The fourth in the series, WPAN Low Rate/ZigBee is the newest and provides specifications for devices that have low data rates, consume very low power and are thus characterized by long battery life. Other standards like Bluetooth and IrDA address high data rate applications such as voice, video and LAN communications.

II. ZigBee Networking Topologies

The network formation is managed by the ZigBee networking layer. The network must be in one of two networking topologies specified in IEEE 802.15.4 star and peer-to-peer.

In the star topology, shown in Figure 6.1 every device in the network can communicate only with the PAN coordinator. A typical scenario in a star network formation is that an FFD programmed to be a PAN coordinator is activated and starts establishing its network. The first thing this PAN coordinator does is select a unique PAN identifier that is not used by any other network in its radio sphere of influence--the region around the device in which its radio can successfully communicate with other radios. In other words, it ensures that the PAN identifier is not used by any other nearby network.

In a peer-to-peer topology(as shown in Figure 6.2), each device can communicate directly with any other device if the devices are placed close enough together to establish a successful communication link. Any FFD in a peer-to-peer network can play the role of the PAN coordinator. One way to decide which device will be the PAN coordinator is to pick the first FFD device that starts communicating as the PAN coordinator. In a peer-to-peer network, all the devices that participate in relaying the messages are FFDs because RFDs are not capable of relaying the messages. However, an RFD can be part of the network and communicate only with one particular device(a coordinator or a router)in the network.

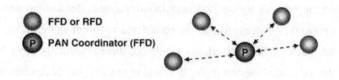

Figure 6.1 A Star Network Topology

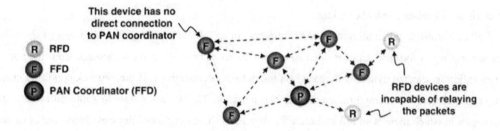

Figure 6.2 A Mesh Networking Topology

A peer-to-peer network can take different shapes by defining restrictions on the devices that can communicate with each other. If there is no restriction, the peer-to-peer network is known as a mesh topology. Another form of peer-to-peer network ZigBee supports is a tree topology (see Figure 6.3). In this case, a ZigBee coordinator (PAN coordinator)establishes the initial network. ZigBee routers form the branches and relay, the messages. ZigBee end devices act as leaves of the tree and do not participate in message routing. ZigBee routers can grow the network beyond the initial network established by the ZigBee coordinator.

Figure 6.3 also shows an example of how relaying a message can help extend the range of the network and even go around barriers. For example, device A needs to send a message to device B, but there is a barrier between them that is hard for the signal to penetrate. The tree topology helps by relaying the message around the barrier and reach device B. This is sometimes referred to as multihopping because a message hops from one node to another until it reaches its destination. This higher coverage comes at the expense of potential high message latency.

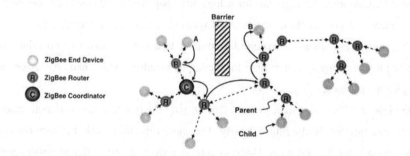

Figure 6.3 A ZigBee Tree Topology

III. **ZigBee Self-Forming and Self-Healing Characteristics**

A ZigBee network starts its formation as soon as devices become active. In a mesh network, for example, the first FFD device that starts communicating can establish itself as the ZigBee coordinator, and other devices then join the network by sending association requests. Because no additional supervision is required to establish a network, ZigBee networks are considered self-forming networks.

On the other hand, when a mesh network is established, there is normally more than one way to relay a message from one device to another. Naturally, the most optimized way is selected to route the message. However, if one of the routers stops functioning due to exhaustion of its battery or if an obstacle blocks the message route, the network can select an alternative route. This is an example of the self-healing characteristic of ZigBee mesh networking.

ZigBee is considered an ad hoc wireless network. In an ad hoc wireless network, some of the wireless nodes are willing to forward data for other devices. The route that will carry a message from the source to the destination is selected dynamically based on the network connectivity. If the network condition changes, it might be necessary to change the routing in the network. This is in contrast to some other networking technologies in which there is an infrastructure in place, and some designated devices always act as routers in the network.

Ⅳ. ZigBee City

In 2004, the Swedish government began drafting new legislation that prompted Göteborg Energi (GE) to start considering different solutions for remote meter reading. At that time, the only developed solutions were based on either power line carrier (PLC) or, to a lesser extent, radio on proprietary or restricted frequencies. The investment cost for a relatively small undertaking of a project of this nature would be substantial, even considering the opportunity to drastically reduce the number of staff involved in manual meter reading. Even if the simplest and cheapest solution were selected, the return on investment would be poor and the overall return would be negative.

However, in 2006, the management team of GE made a bold decision. Forced by new legislation to implement advanced metering infrastructure (AMI), the team decided to examine, in detail, existing and new technologies available for automated meter reading. The team had to make a strategic decision: either adopt the cheapest possible system just to adhere to the new legislation, or seek greater savings and extra revenue to warrant investment in a more advanced system. The older proven technologies did not meet the performance levels demanded. Furthermore, the potential for a long life span was diminished. GE decided to make the most of its investment and look for the most advanced system available at a reasonable cost.

It was a brave decision when GE awarded a contract to a supplier that had no prior installations in Europe, and to prime, the company responsible for the implementation, which had no previous experience in undertaking such a huge project.

It was the view of GE's senior management that, to reap the real benefits of remote reading, readings must not only be monthly but, in the future, hourly. The shorter the time cycle between readings, the more useful the information is for the end users. Hourly readings provide a very different understanding of usage pattern than monthly or yearly readings. The more advanced metering systems also offered a much broader spectrum of possible services than those previously used. Hourly readings, on-demand readings in real time, remote connect/disconnect, power-failure alarm in real time, monitoring of power usage and voltage levels, and other advanced functionalities were among the features GE management anticipated.

The tendering process was initiated, and GE had received eight tenders by the end of 2006. The solutions offered covered all available means of communication available on the market as well as one new solution. Four suppliers supported PLC as the main form of communication, two had selected general packet radio service (GPRS), one suggested radio on a restricted frequency and one, the new entrant on the European market, had ZigBee as the main form of communication.

Internally, the project focus prior to the tendering procedure had been centered on which form of communication was best, the cheaper PLC or the more advanced GPRS. During the evaluation of the tenders, and in light of some of the problems other utilities had experienced with their chosen solutions, it became clear the key was really the central metering system at the heart of the different solutions.

Some of the systems were very basic and some were not beyond the design stage. The project team was surprised when one supplier offered the opportunity to become the first customer of its new system under development, but it was not the wish of the project team to become a guinea pig in developing a system from scratch.

Finally, it became clear one supplier had a system that was beyond the others in functionality, user friendliness and advanced thinking. The metering infrastructure was based on ZigBee as the main form of communication, which was very new at the time. There were no major ZigBee implementations for AMI in the world. The fact ZigBee was an open standard and the possibilities this presented were considerably alluring. The supplier also was able to present a smaller installation that used the system for gas and water metering, as well. This proved the system could handle multi-metering. The ZigBee system satisfied the GE objectives, and the utility decided to make the most of the investment by selecting the most advanced system available at a reasonable cost.

In January 2008, GE undertook a pilot to prepare for the actual rollout. All the processes, hardware, software, integrations, staff involved and suppliers performed a dress rehearsal. This provided a thorough check of what needed to be corrected before full-scale implementation.

In March 2008, the large installation phase started. For a period of more than 15 months, approximately 70 electricians installed up to 1,500 meters per day. Appointments were planned in advance to access the meters of 42,000 customers. A separate call center was set up to handle these bookings and respond to questions regarding the installations. An extensive program for customer communication was undertaken to ensure all customers were informed about the project, the reason for it and the benefits for the end customers.

The most complex part of the project was the necessary data handling. A huge amount of information had to be collected from various systems on the customer side, cleaned and aggregated, and transferred to prime in a controlled manner. The information created at each installation (for example, new meter identification, metering value and time for installation) had to be transferred back to the customer and then distributed in a timely manner to various systems such as the customer information system, billing system, meter data management system and so forth. To effectively handle all data, an installation system was created based on the business model for the rollout. This system made it possible to keep track of each and every individual installation, see when it was done, by whom and whether all the data was collected correctly, and when the meter installation was approved based on a set of predefined criteria.

By January 2010, more than 265,000 meters had been installed, the performance level for monthly reads was approximately 99.8%, 8,000 concentrators had been installed and the project entered the final clean-up phase. A couple hundred meters are not yet installed because of problems accessing the customer site. Of the installed meters, approximately 500 meters still need some work to communicate regularly with the concentrators. In some cases, they need an extra antenna, and in other cases, there are software issues or other problems.

V. New Words

current ['kʌr(ə)nt]　　　　　　　　n. 电流；（水，气，电）流
parameter [pə'ræmɪtə]　　　　　　n. 参数；系数；参量
formulate ['fɔːmjʊlet]　　　　　　 v. 制定；用公式表示；明确地表达
topology [tə'pɒlədʒɪ]　　　　　　n. 拓扑结构；n. 拓扑学；地志学
coordinator [kəʊ'ɔːdɪneɪtə]　　　n. 协调器；协调者；[自] 协调器

VI. Phrases

micro controllers　　　　　　　微控制器
radio bands　　　　　　　　　　电频段
network layer　　　　　　　　　网络层
tree networks　　　　　　　　　树形网络
coordinator device　　　　　　　协调器设备
the central node　　　　　　　　中央节点
the network level　　　　　　　网络级
networking layer　　　　　　　　网络层；网路层
communication link　　　　　　　通信链路
self-forming networks　　　　　　自组织网络
the network connectivity　　　　　网络连通性

VII. Abbreviations

PAN　　Personal Area Network　　个域网
FFD　　Fully Functional Devices　　全功能设备

Passage B IEEE 802.15.4 Technology and 6LoWPAN

Ⅰ. Beacon Frame

Figure 6.4 shows the structure of the beacon frame, which originates from within the MAC sublayer. A coordinator can transmit network beacons in a beacon-enabled PAN. The MAC payload contains the superframe specification, GTS fields, pending address fields, and beacon payload. The MAC payload is prefixed with a MAC header (MHR) and appended with a MAC footer (MFR). The MHR contains the MAC Frame Control field, beacon sequence number (BSN), addressing fields, and optionally the auxiliary security header. The MFR contains a 16-bit frame check sequence (FCS). The MHR, MAC payload, and MFR together form the MAC beacon frame (i.e., MPDU).

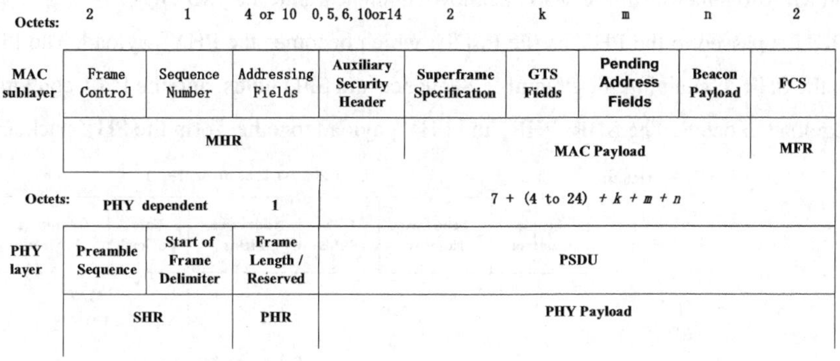

Octets:	2	1	4 or 10	0, 5, 6, 10 or 14	2	k	m	n	2
MAC sublayer	Frame Control	Sequence Number	Addressing Fields	Auxiliary Security Header	Superframe Specification	GTS Fields	Pending Address Fields	Beacon Payload	FCS
	MHR				MAC Payload				MFR

Octets:	PHY dependent	1	7 + (4 to 24) + k + m + n
PHY layer	Preamble Sequence	Start of Frame Delimiter	Frame Length / Reserved	PSDU
	SHR		PHR	PHY Payload

Figure 6.4 Schematic view of the beacon frame and the PHY packet

The MAC beacon frame is then passed to the PHY as the PHY service data unit (PSDU), which becomes the PHY payload. The PHY payload is prefixed with a synchronization header (SHR), containing the Preamble Sequence and Start-of-Frame Delimiter (SFD) fields, and a PHY header (PHR) containing the length of the PHY payload in octets. The SHR, PHR, and PHY payload together form the PHY packet (i.e., PPDU).

Ⅱ. Data Frame

Figure 6.5 shows the structure of the data frame, which originates from the upper layers.

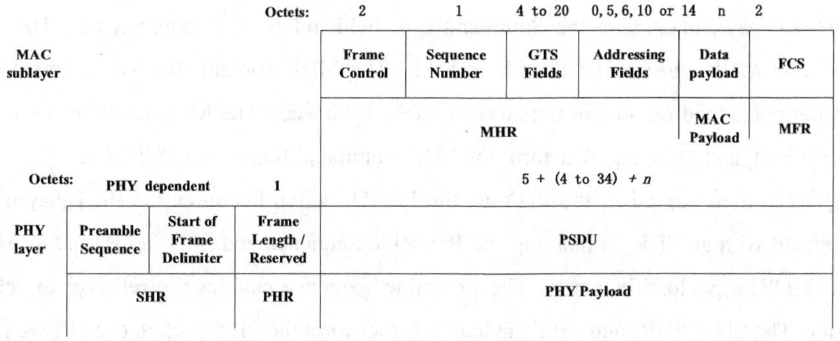

Octets:	2	1	4 to 20	0, 5, 6, 10 or 14	n	2
MAC sublayer	Frame Control	Sequence Number	GTS Fields	Addressing Fields	Data payload	FCS
	MHR				MAC Payload	MFR

Octets:	PHY dependent	1	5 + (4 to 34) + n
PHY layer	Preamble Sequence	Start of Frame Delimiter	Frame Length / Reserved	PSDU
	SHR		PHR	PHY Payload

Figure 6.5 Schematic view of the data frame and the PHY packet

The data payload is passed to the MAC sublayer and is referred to as the MAC service data unit (MSDU). The MAC payload is prefixed with an MHR and appended with an MFR. The MHR contains the Frame Control field, data sequence number (DSN), addressing fields, and optionally the auxiliary security header. The MFR is composed of a 16-bit FCS. The MHR, MAC payload, and MFR together form the MAC data frame, (i.e., MPDU).

The MPDU is passed to the PHY as the PSDU, which becomes the PHY payload. The PHY payload is prefixed with an SHR, containing the Preamble Sequence and SFD fields, and a PHR containing the length of the PHY payload in octets. The preamble sequence and the data SFD enable the receiver to achieve symbol synchronization. The SHR, PHR, and PHY payload together form the PHY packet, (i.e., PPDU).

III. Acknowledgment Frame

Figure 6.6 shows the structure of the acknowledgment frame, which originates from within the MAC sublayer. The MAC acknowledgment frame is constructed from an MHR and an MFR; it has no MAC payload. The MHR contains the MAC Frame Control field and DSN. The MFR is composed of a 16-bit FCS. The MHR and MFR together form the MAC acknowledgment frame (i.e., MPDU).

The MPDU is passed to the PHY as the PSDU, which becomes the PHY payload. The PHY payload is prefixed with the SHR, containing the Preamble Sequence and SFD fields, and the PHR containing the length of the PHY payload in octets. The SHR, PHR, and PHY payload together form the PHY packet, (i.e.,PPDU).

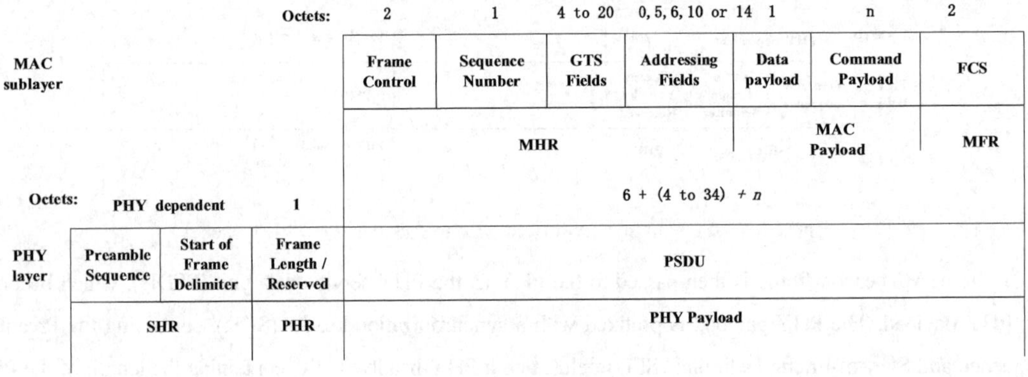

Figure 6.6 Schematic view of the acknowledgment frame and the PHY packet

IV. MAC Command Frame

Figure 6.7 shows the structure of the MAC command frame, which originates from within the MAC sublayer. The MAC payload contains the Command Type field and the command payload. The MAC payload is prefixed with an MHR and appended with an MFR. The MHR contains the MAC Frame Control field, DSN, addressing fields, and optionally the auxiliary security header. The MFR contains a 16- bit FCS. The MHR, MAC payload, and MFR together form the MAC command frame, (i.e., MPDU).

The MPDU is then passed to the PHY as the PSDU, which becomes the PHY payload. The PHY payload is prefixed with an SHR, containing the Preamble Sequence and SFD fields, and a PHR containing the length of the PHY payload in octets. The preamble sequence enables the receiver to achieve symbol synchronization. The SHR, PHR, and PHY payload together form the PHY packet, (i.e., PPDU).

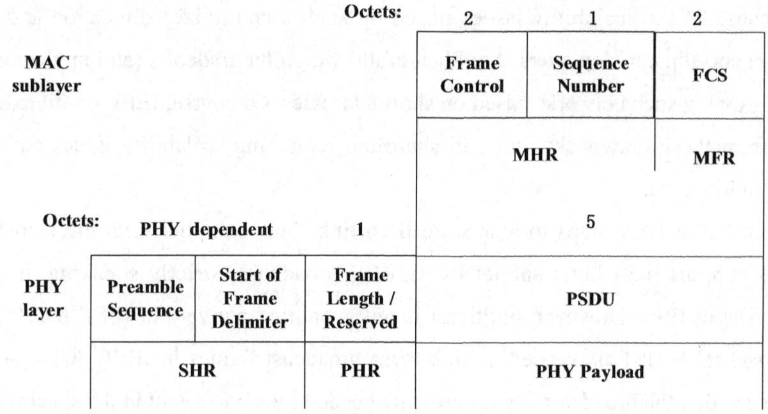

	Octets:	2	1	2
MAC sublayer		Frame Control	Sequence Number	FCS
		MHR		MFR

	Octets:	PHY dependent		1	5
PHY layer		Preamble Sequence	Start of Frame Delimiter	Frame Length / Reserved	PSDU
		SHR		PHR	PHY Payload

Figure 6.7 Schematic view of the MAC command frame and the PHY packet

V. IEEE 802.15.4 Mode for IP

IEEE 802.15.4 defines four types of frames: beacon frames, MAC command frames, acknowledgement frames, and data frames. IPv6 packets MUST be carried on data frames. Data frames may optionally request that they be acknowledged. In keeping with [IETF RFC3819], it is recommended that IPv6 packets be carried in frames for which acknowledgements are requested so as to aid link-layer recovery.

IEEE 802.15.4 networks can either be nonbeacon-enabled or beacon-enabled [ieee802.15.4]. The latter is an optional mode in which devices are synchronized by a so-called coordinator's beacons. This allows the use of superframes within which a contention-free Guaranteed Time Service (GTS) is possible. This document does not require that IEEE networks run in beacon-enabled mode. In nonbeacon- enabled networks, data frames (including those carrying IPv6 packets) are sent via the contention-based channel access method of unslotted CSMA/CA.

In nonbeacon-enabled networks, beacons are not used for synchronization. However, they are still useful for link-layer device discovery to aid in association and disassociation events. This document recommends that beacons be configured so as to aid these functions. A further recommendation is for these events to be available at the IPv6 layer to aid in detecting network attachment, a problem being worked on at the IETF at the time of this writing.

The specification allows for frames in which either the source or destination addresses (or both) are elided. The mechanisms require that both source and destination addresses be included in the IEEE 802.15.4 frame header. The source or destination PAN ID fields may also be included.

IEEE 802.15.4 defines several addressing modes: it allows the use of either IEEE 64-bit extended addresses or (after an association event) 16-bit addresses unique within the PAN [IEEE802.15.4]. This document supports both 64-bit extended addresses, and 16-bit short addresses.

For use within 6LoWPANs, this document imposes additional constraints (beyond those imposed by IEEE 802.15.4) on the format of the 16-bit short addresses, as specified in Section 12. Short addresses being transient in nature, a word of caution is in order: since they are doled out by the PAN coordinator function during an association event, their validity and uniqueness is limited by the lifetime of that association. This can be cut short by the expiration of the association or simply by any mishap occurring to the PAN coordinator. Because of the scalability issues posed by such a centralized allocation and single point of failure at the PAN coordinator, deployers should carefully weigh the tradeoffs (and implement the necessary mechanisms) of growing such networks based on short addresses. Of course, IEEE 64-bit extended addresses may not suffer from these drawbacks, but still share the remaining scalability issues concerning routing, discovery, configuration, etc.

It is assumed that a PAN maps to a specific IPv6 link. This complies with the recommendation that shared networks support link- layer subnet [RFC3819] broadcast. Strictly speaking, it is multicast not broadcast that exists in IPv6. However, multicast is not supported natively in IEEE 802.15.4. Hence, IPv6 level multicast packets MUST be carried as link-layer broadcast frames in IEEE 802.15.4 networks. This MUST be done such that the broadcast frames are only heeded by devices within the specific PAN of the link in question.

The MTU size for IPv6 packets over IEEE 802.15.4 is 1280 octets. However, a full IPv6 packet does not fit in an IEEE 802.15.4 frame. 802.15.4 protocol data units have different sizes depending on how much overhead is present [IEEE802.15.4]. Starting from a maximum physical layer packet size of 127 octets (a Max PHY Packet Size) and a maximum frame overhead of 25 (a Max Frame Overhead), the resultant maximum frame size at the media access control layer is 102 octets.

Link-layer security imposes further overhead, which in the maximum case (21 octets of overhead in the AES-CCM-128 case, versus 9 and 13 for AES-CCM-32 and AES-CCM-64, respectively) leaves only 81 octets available. This is obviously far below the minimum IPv6 packet size of 1280 octets, and in keeping with Section 5 of the IPv6 specification [RFC2460], a fragmention and reassembly adaptation layer must be provided at the layer below IP.

Furthermore, since the IPv6 header is 40 octets long, this leaves only 41 octets for upper-layer protocols, like UDP. The latter uses 8 octets in the header which leaves only 33 octets for application data. Additionally, as pointed out above, there is a need for a fragmentation and reassembly layer, which will use even more octets.

VI. LoWPAN Adaptation Layer and Frame Format

The encapsulation formats defined in this section (subsequently referred to as the "LoWPAN encapsulation") are the payload in the IEEE 802.15.4 MAC protocol data unit (PDU). The LoWPAN payload (e.g., an IPv6 packet) follows this encapsulation header.

All LoWPAN encapsulated datagrams transported over IEEE 802.15.4 are prefixed by an encapsulation header stack. Each header in the header stack contains a header type followed by zero or more header fields.

Whereas in an IPv6 header the stack would contain, in the following order, addressing, hop-by-hop options, routing, fragmentation, destination options, and finally payload [RFC2460]; in a LoWPAN header, the analogous header sequence is mesh (L2) addressing, hop-by-hop options (including L2 broadcast/multicast), fragmentation, and finally payload. These examples show typical header stacks that may be used in a LoWPAN network.

A LoWPAN encapsulated IPv6 datagram:

```
+---------------+-------------+---------+
| IPv6 Dispatch | IPv6 Header | Payload |
+---------------+-------------+---------+
```

A LoWPAN encapsulated LOWPAN_HC1 compressed IPv6 datagram:

```
+--------------+------------+---------+
| HC1 Dispatch | HC1 Header | Payload |
+--------------+------------+---------+
```

A LoWPAN encapsulated LOWPAN_HC1 compressed IPv6 datagram that requires mesh addressing:

```
+-----------+-------------+--------------+------------+---------+
| Mesh Type | Mesh Header | HC1 Dispatch | HC1 Header | Payload |
+-----------+-------------+--------------+------------+---------+
```

A LoWPAN encapsulated LOWPAN_HC1 compressed IPv6 datagram that requires fragmentation:

```
+-----------+-------------+--------------+------------+---------+
| Frag Type | Frag Header | HC1 Dispatch | HC1 Header | Payload |
+-----------+-------------+--------------+------------+---------+
```

A LoWPAN encapsulated LOWPAN_HC1 compressed IPv6 datagram that requires both mesh addressing and fragmentation:

```
+-------+-------+-------+-------+---------+---------+---------+
| M Typ | M Hdr | F Typ | F Hdr | HC1 Dsp | HC1 Hdr | Payload |
+-------+-------+-------+-------+---------+---------+---------+
```

A LoWPAN encapsulated LOWPAN_HC1 compressed IPv6 datagram that requires both mesh addressing and a broadcast header to support mesh broadcast/multicast:

```
+-------+-------+-------+-------+---------+---------+---------+
| M Typ | M Hdr | B Dsp | B Hdr | HC1 Dsp | HC1 Hdr | Payload |
+-------+-------+-------+-------+---------+---------+---------+
```

When more than one LoWPAN header is used in the same packet, they MUST appear in the following order:
- Mesh Addressing Header
- Broadcast Header
- Fragmentation Header

All protocol datagrams (e.g., IPv6, compressed IPv6 headers, etc.) SHALL be preceded by one of the valid LoWPAN encapsulation headers, examples of which are given above. This permits uniform software treatment of datagrams without regard to the mode of their transmission.

The definition of LoWPAN headers, other than mesh addressing and fragmentation, consists of the dispatch value, the definition of the header fields that follow, and their ordering constraints relative to all other headers. Although the header stack structure provides a mechanism to address future demands on the LoWPAN adaptation layer, it is not intended to provided general purpose extensibility. This format document specifies a small set of header types using the header stack for clarity, compactness, and orthogonality.

- Dispatch: 6-bit selector. Identifies the type of header immediately following the Dispatch Header.
- Type-specific header: A header determined by the Dispatch Header.

The dispatch value may be treated as an unstructured namespace. Only a few symbols are required to represent current LoWPAN functionality. Although some additional savings could be achieved by encoding additional functionality into the dispatch byte, these measures would tend to constrain the ability to address future alternatives. The dispatch value bit pattern is shown as Figure 6.8.

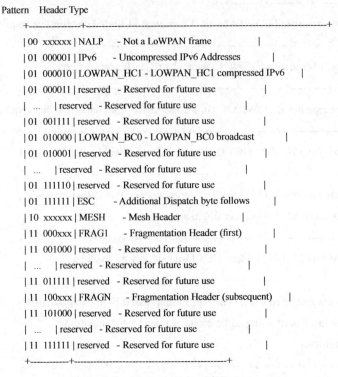

```
             Pattern    Header Type
            +------------+---------------------------------------------------+
            | 00 xxxxxx  | NALP        - Not a LoWPAN frame                  |
            | 01 000001  | IPv6        - Uncompressed IPv6 Addresses         |
            | 01 000010  | LOWPAN_HC1  - LOWPAN_HC1 compressed IPv6          |
            | 01 000011  | reserved    - Reserved for future use             |
            |     ...    | reserved    - Reserved for future use             |
            | 01 001111  | reserved    - Reserved for future use             |
            | 01 010000  | LOWPAN_BC0  - LOWPAN_BC0 broadcast                |
            | 01 010001  | reserved    - Reserved for future use             |
            |     ...    | reserved    - Reserved for future use             |
            | 01 111110  | reserved    - Reserved for future use             |
            | 01 111111  | ESC         - Additional Dispatch byte follows    |
            | 10 xxxxxx  | MESH        - Mesh Header                         |
            | 11 000xxx  | FRAG1       - Fragmentation Header (first)        |
            | 11 001000  | reserved    - Reserved for future use             |
            |     ...    | reserved    - Reserved for future use             |
            | 11 011111  | reserved    - Reserved for future use             |
            | 11 100xxx  | FRAGN       - Fragmentation Header (subsequent)   |
            | 11 101000  | reserved    - Reserved for future use             |
            |     ...    | reserved    - Reserved for future use             |
            | 11 111111  | reserved    - Reserved for future use             |
            +------------+---------------------------------------------------+
```

Figure 6.8 Dispatch Value Bit Pattern

- NALP: Specifies that the following bits are not a part of the LoWPAN encapsulation, and any LoWPAN node that encounters a dispatch value of 00xxxxxx shall discard the packet. Other non-LoWPAN protocols that wish to coexist with LoWPAN nodes should include a byte matching this pattern immediately following the 802.15.4 header.
- IPv6: Specifies that the following header is an uncompressed IPv6 header [RFC2460].
- LOWPAN_HC1: Specifies that the following header is a LOWPAN_HC1 compressed IPv6 header.
- LOWPAN_BC0: Specifies that the following header is a LOWPAN_BC0 header for mesh broadcast/multicast support..
- ESC: Specifies that the following header is a single 8-bit field for the Dispatch value. It allows support for Dispatch values larger than 127.

Ⅶ. New Words

beacon ['biːkən]　　　　　　　　　n. 信标

frame [freɪm]　　　　　　　　　　n. 帧；框架

sublayer ['sʌbˌleɪə]　　　　　　　n. 子层；下[低，次，内]层，亚表层

coordinator [kəʊ'ɔːdɪneɪtə]　　　n. 协调器；同等的人（或物）

superframe ['sjuːpəfreɪm]　　　　n. 超帧

packet ['pækɪt]　　　　　　　　　n. 包；信息包

octet [ɒk'tet]		n. 八位位组，八位字节
encapsulation [ɪnˈkæpsəˈleɪʃən]		n. 封装；包装
datagrams ['detə'græm]		n. 数据报
prefixed [priː'fikst]		adj. 有前缀的
orthogonality [ˈɔːθɒgəˈnælətɪ]		n. 正交性；相互垂直

Ⅷ. Phrases

beacon-enabled	信标使能
preamble sequence	前导序列
data frame	数据帧
addressing fields	地址域
acknowledgment frame	确认帧
Hop-by-Hop	跳路由；逐跳

Ⅸ. Abbreviations

MAC	Media Access Control	媒体访问控制
PAN	Personal Area Network	个域网
GTS	Geostationary Technology Satellite	地球同步技术卫星
MHR	MAC header	帧头
MFR	MAC footer	帧尾
BSN	beacon sequence number	信标序列号
FCS	frame check sequence	帧校验序列
PSDU	PHY service data unit	服务数据单元
SHR	synchronization header	同步头
SFD	Start-of-Frame Delimiter	帧首定界符
PHR	PHY header	PHY 报头
MSDU	MAC service data unit	MAC 服务数据单元
DSN	data sequence number	数据序列号
AES	Advanced Encryption Scheme	先进加密算法
CSMA/CA	Carrier Sense Multiple Access/Collision Avoidance	载波侦听冲突避免
FFD	Full Function Device	全功能设备
GTS	Guaranteed Time Service	保障时间服务
MTU	Maximum Transmission Unit	最大传输单元
NALP	Not a LoWPAN frame	非 LoWPan 帧

Passage C Online Monitoring System for Transmission Lines and Sensing of Traffic Flows

Ⅰ. Online Monitoring System for Transmission Lines

The condition of transmission lines are directly affected by wind, rain, snow, fog, ice, lightning and other natural forces; at the same time industrial and agricultural pollution are also a direct threat to the safe operation of transmission lines. The operating environment of transmission lines and the operating states are very complex, which requires more automatic monitoring, more control and protection equipment to automatically send alarms when accidents occur and dispatch strategy adjustment according to the operation mode, thus the faults will be processed at the early phase or be isolated in a small range.

Traditional wired communications cannot meet the communication needs of online monitoring of transmission lines. WSNs have an advantage of the strong ability to adapt to harsh environments, large area coverage, self-organization, self-configuretion guration and strong utility independence and are very suitable for data communication monitoring systems for transmission lines.

With the technical advantages of WSNs, establishing a full range, multielement online monitoring system can send timely warnings of disasters, rapidly locate the positions of faults, sense transmission line faults, shorten the time of fault recovery and thus improve the reliability of the power supply. WSN use can not only effectively prevent and reduce power equipment accidents, when combining with the conductor temperature, environmental and meteorological real-time online monitoring, but can also provide data to support transmission efficiency improving and increasing dynamic capacity for transmission lines.

The general architecture of online monitoring systems for transmission lines is shown in Figure 6.9. Currently, some provincial power companies of the State Grid Corporation of China (SGCC) are promoting applications of WSN technology in the online monitoring of transmission line. For example, since 2013, Liaoning and Ningxia electric power companies are developing demonstration projects based on WSN for transmission lines online monitoring system.

Ⅱ. Sensing of Traffic Flows

Intelligent traffic management solutions rely on the accurate measurement and reliable prediction of traffic flows within a city. This includes not only an estimation of the density of cars on a given street or the number of passengers inside a given bus or train but also the analysis of the origins and destinations of the vehicles and passengers.

Monitoring the traffic situation on a street or intersection can be achieved by means of traditional wired sensors, such as cameras, inductive loops, etc. While wireless technology can be beneficial in reducing deployment costs of such sensors, it does not directly affect the accuracy or usefulness of the measurement results.

However, by broadening the definition of the term "sensor" and making use of wireless technology readily available in many vehicles and smart phones, the vehicles themselves as well as the passengers using the public transportation systems can become "sensors" for the accurate measurement of traffic flows within a city.

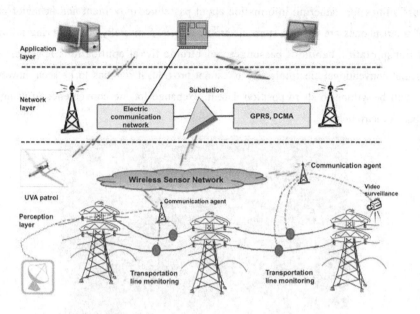

Figure 6.9　General architecture of online monitoring system for transmission line based on WSNs

Techniques for collecting traffic flow data from vehicles are collectively referred to as floating car data (FCD). This includes methods relying on a relatively small number of vehicles explicitly transmitting their position information to a central server (e.g. taxis or buses sending their position obtained via GPS) as well as approaches relying on location information of mobile phones obtained from real-time location databases of the cellular network operators. The latter approach does not actually involve any sensing by the vehicle itself, but still makes use of a wireless network (i.e. the existing cellular network) to sense or rather infer the current characteristics of traffic flows. The technical challenges lie particularly in the processing of the potentially large amounts of data, the distinction between useful and non-useful data and the extrapolation of the actual traffic flow data from the observation of only a subset of all vehicles.

Extensions of the FCD idea involving information gathered from the on-board electronics of the vehicles have been proposed under the term extended floating car data (XFCD). Collecting and evaluating data from temperature sensors, rain sensors, ABS, ESC and traction control system of even a relatively small number of cars can be used to derive real-time information about road conditions which can be made available to the public and/or used for an improved prediction of traffic flows based on anticipated behavior of drivers in response to the road conditions.

Privacy issues must be taken into consideration whenever location or sensor data is collected from private vehicles. However, this is a general concern related to the monitoring of traffic flows, and schemes that don't make use of wireless technology (e.g. relying on license-plate recognition) also have to consider the car owners' privacy.

Equivalent to the measurement of vehicle movement by FCD, passenger behavior in public transportation systems can be analyzed with the help of wireless technology. For example electronic tickets (as shown in Figure 6.10), which typically employ RFID technology for registering the access to a subway station, bus or tram, effectively turn the passenger into a part of a sensor network.

The possibilities for gathering information about passenger movement and behavior can be further increased if smart phones are used to store electronic tickets. Especially for gathering information about intermodal transportation habits of passengers, electronic ticket applications for smart phones offer possibilities that conventional electronic tickets cannot provide. It remains to be seen, however, to which extent users will be willing to share position data in exchange for the convenience of using their mobile phone as a bus or metro ticket.

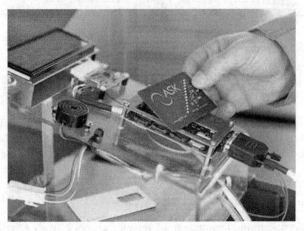

Figure 6.10　Electronic tickets for smarter travel

Ⅲ. **New Words**

self-organization ['self,ɔ:gənaɪ'zeɪʃən]　*n.* 自组织

Ⅳ. **Phrases**

automatic monitoring	自动监控
self-configuration	自配置
cellular network	蜂窝网
license-plate recognition	车牌识别

Ⅴ. **Abbreviations**

XFCD　extended floating car data　　　　　　扩展浮动车数据

Exercises

Translate the following sentences into Chinese or English.

1. M2M, gaining popularity among consumers and service providers to improve the lives of people, is an approach where devices such as sensors and actuators connected with each other or with a computer system for data analysis for producing immediate action.

2. Low computing power of the ZigBee's sensor node is causing low endurance against network security attacks, thus providing solutions to all types of network security attacks which are likely to cause significant overhead, so it is vital to properly rank security attacks by evaluating their severity and selecting appropriate countermeasures.

3. 估算精确的到达时间（TOA）是在射频定位系统中使用无线传感器网络（WSN）的最重要的技术之一，因为估算到达时间（TOA）的精度是与射频信号带宽成比例的，所以使用宽频带是实现高精度的最基本的方法。

4. 无线传感器网络（WSN）是一项空间分布的自主传感器来感知特别任务，就像无线个域网网络形成的互连形式简单、低功耗、低处理能力的无线设备。

参考译文 Passage A ZigBee技术

Ⅰ. 概述

ZigBee是一种低成本、低功耗的无线网状网络标准，旨在延长无线控制和监控应用设备的生命周期。ZigBee设备具有低延迟性，这进一步降低了它的平均电流。ZigBee芯片通常集成了射频芯片以及具有60-256KB闪存的微控制器。ZigBee可用于工业、科学和医学（ISM）中的无线电频段：世界大多数使用的频段范围是2.4G，中国是784MHz，欧洲是868MHz，美国和澳大利亚则是915MHz。数据传输速率在20kbit/s（868MHz频段）到250kbit/s（2.4GHz频段）范围之间。

ZigBee网络层不仅支持星型和树型网络，还支持通用的网状网络。每种网络中必须具备一个协调器设备，此设备主要用于创建、控制和维护参数。在星型网络中，协调器必须是中央节点。树型网络和网状网络都允许用ZigBee路由来扩展网络级中的通信。

ZigBee是由IEEE 802.15工作组下面一个重要的工作小组制定的，它是全球通信协议标准之一。作为系列中的第4种，低速无线个域网/ZigBee是最新的标准，并且为低数据传输率、低功耗设备提供了规范，所以具有电池寿命长的特点。蓝牙和红外等其他标准强调高数据传输速率，应用在音频、视频和局域网通信等方面。

Ⅱ. ZigBee网络拓扑结构

ZigBee网络层管理网络构造。该网络必须是IEEE802.15.4规定的两种网络拓扑结构中的一种，即星型网络或对等网络。

图6.1所示的星型网络中，每个设备都只与局域网中的协调器进行通信。星型网络中，典型的方案是，被程序化为局域网协调器的全功能设备被激活，而且开始组网。此局域网协调器首先做的是，选择一个未在其他通信范围内使用的独特的域网标识符。设备的通信范围是指，设备通过无线电能够成功地与其他设备通信的范围。换句话说，域网标识符必须保证没有被附近的其他网络使用。

图6.2所示的对等网络中，设备都能直接与其他设备通信，只要此设备被放置于足够近的位置以成功建立通信链路。任意一个对等网络中的全功能设备（FFD）都可以作为局域网的协调器。决定哪个设备将作为局域网协调器的方法是，选择通信的第一个全功能设备作为局域网的协调器。在对等网络中，参与转发消息的设备是全功能设备，因为精简功能设备不能转发消息。然而，精简功能设备可以作为网络的一部分，它只与网络中特定的设备（协调器或者路由器）通信。

通过确定设备相互通信的限制，对等网络可以采取不同的形状。如果没有任何限制，对等网络就是树形网络（如图6.3所示）。在这种情况下，ZigBee协调器（局域网协调器）建立起初始网络。ZigBee路由器的功能是组件分支和转发消息，ZigBee终端设备作为叶子节点不参与信息路由选择。ZigBee路由器可以在ZigBee协调器建立的初始网络上扩展网络。

图6.3也举例说明了如何转发消息可以扩展网络范围甚至绕过障碍物。例如，A设备要发送一条消息给B设备，但它们之间有阻挡信号穿透的障碍物。树型网络转发消息时会绕过障碍物到达B设备。有时这被称为多跳，因为消息从一个节点经过多跳，从而达到目的节点。这种高覆盖率解决了潜在的高消息延迟。

Ⅲ. 无线个域网的形成和自愈特征

当设备被激活时，ZigBee网络便开始形成。例如，在网状网络中，首先形成的全功能（FFD）设备开始通信能使它自身成为ZigBee协调器，然后其他设备通过发送连接请求加入网络。因为建立网络不需要额外的管理，所以ZigBee网络被认为是自组织网络。

另一方面，当一个网状网络建立之后，从一个设备到另一个设备通常会有多条路径转发消息。显然，最优路径会被选来发送消息。但如果其中一个路由因电能耗尽而停止运行，或者因障碍物阻塞了消息路由，网络就可以选择备用路由。这是ZigBee网状网络自愈特性的实例。

ZigBee被认为是一种ad hoc无线网。在ad hoc无线网络中，一些无线节点自动向其他设备转发数据。基于网络的连通性，从源节点到目的节点传输消息的路径可动态选取。如果网络环境变化，可能需要改变网络中的路由配置。这与一些其他网络技术形成鲜明对比，因为这些网络技术中有基础设施，而且一些指定的设备在网络中扮演路由的角色。

Ⅳ. ZigBee城市

2004年，瑞典政府开始起草新的法律，激励哥德堡能源公司（GE）开始探索远程抄表的不同解决方案。在那时，相对成熟的解决方案要么基于电力线载波，要么在较小程度上基于专用无线信道或限制频率。尽管，可以大幅削减人工抄表的员工数量，但对于相对较小的项目，会有巨大的投资。即使选择最简单、最廉价的解决方案，投资回报也会很少而导致总体的收益为负。

然而，GE的管理团队在2006年做出了一个大胆的决定，迫于新法律要求实施高级计量体系（AMI），该团队决定详细调查现有的，可用于自动抄表的新技术。该团队务必做出战略性的决定：要么为了支持用户新法规而采用最便宜的系统，要么寻求更多的资金以确保投资更先进的系统。已有证明，旧技术不能满足性能等级的要求。此外，生命周期长的可能性也降低了。GE决定投入大量资金来寻求合理成本条件下最先进的系统。

GE做了一个大胆的决定，与一家在欧洲从没有过安装经验的供应商签署了合同，该公司首要的责任是，在毫无经验的情况下，处理如此巨大项目。

GE的高级管理人员认为，在以后，只有按小时抄表而不是按月抄表才能实现远程抄表的益处。抄表的时间周期越短，读取的信息就对终端用户越有用。相比按月抄表或者按年抄表，按小时抄表提供了一种不同的使用模式。同时，高级计量系统也比之前的系统有更大的可能性服务范围。按小时抄表、实时按需抄表、远程连接/断开、实时断电报警、电量使用监控和电压水平，以及其他高级功能都是GE管理团队所预期的特性。

招标程序截止到2006年，GE收到了8家竞标者。提供的解决方案涵盖了当时市场上所有的可能通信方式，也有一种新的解决方案。有4家供应商支持采用电力线载波作为主要的通信方式，2家供应商选择通用无线分组业务（GPRS），一家供应商建议使用受限频率的无线电，还有一家新进军欧洲市场的公司采用Zigbee作为主要的通信方式。

在GE内部，该项目主要关注招标过程集中选择最好的通信方式，更廉价的电力线载波，更先进的GPRS。据其他公用事业的解决方案的经历，评估竞标者的过程中，显而易见，中央计量系统处于不同解决方案的核心。

一些设计系统非常基础，并且有一些都没有超越设计阶段。供应商提供机会，并成为他们正在开发的新系统的第一个消费者时，会使项目团队非常惊奇，但项目团队并不希望成为一个新系统的实验品。

最后，很明显的是，供应商拥有的系统在功能上超过了其他系统，并且拥有用户友好和先进思考的功能。基于ZigBee计量基础架构，作为主要的交流方式，其构思非常新颖。世界上还没有任何公司实现针对AMI的主要无线个域网。事实上，无线个域网是个开放的标准，并且其发展潜力也非常吸引人。供应商也可以提供一个较小的装置利用于天然气和水计量系统。无线个域网满足了GE目标，通过选择大量价格合理的先进系统，其实效性带来了大量的投资。

2008年1月，GE进行了试点，为实际推广做准备。所有的流程、硬件、软件、集成、员工都包含在内，供应商还进行了彩排。这提供了彻底的检查，用于全面实现之前需要检查的部分。

2008年3月，开始大型的安装阶段。这一时期超过15个月，大约每天70位电工安装1500m。提前预约了42,000名客户的访问计划。独立的呼叫中心用于处理这些关于安装的预订，解答相关问题。制定一个面向大规模客户的沟通程序确保顾客能够了解该产品，包括该产品的生产动机，以及客户最终能得到的利益。

项目最复杂的部分是必要的数据处理。大量的信息收集来自于客户端的各类系统，清理、聚合以及转移为主要的控制方式。每个安装创建的信息（例如，新仪表识别、计量值以及安装时间）都会转移回客户，然后及时分布给各类系统，例如用户信息系统、计费系统、仪表数据管理系统等。为有效处理所有的数据、安装系统都基于推广的商业模式。此系统可以记录单个的安装，可以看到何时工作、由谁操作、是否所有的数据都是正确收集，以及基于预定义标准的仪表装置何时得到认可。

到2010年1月，安装已经超过265000个仪表，其月读的性能水平大约达到99.8%，8000个集中器已经被安装，并且项目已经进入收尾阶段。由于访问客户网站的问题，还有几百米没有安装好。大约有500m已经安装好了的仪表，仍然需要定期与用户进行交流。在某些情况下，他们需要额外天线，还有一些情况，会存在软件或其他方面的问题。

参考答案

1. 机对机（M2M）是一种使设备相连接的方式，如传感器和执行器相互连接或通过计算机系统分析数据来产生即时操作，它改善了人们的生活，深受消费者和服务供应商的欢迎。

2. 无线个域网（ZigBee）传感器节点计算能力低是造成网络安全攻击防御能力低的原因，从而为所有网络安全攻击类型提供解决方案可能造成巨额开销。因此，通过对其严重性进行评估，选择合适的对策，来适当地将安全攻击分级至关重要。

3. The precise time-of-arrival (TOA) estimation is one of the most important techniques in RF-based positioning systems that use wireless sensor networks (WSNs), and because the accuracy of TOA estimation is proportional to the RF signal bandwidth, and using broad bandwidth is the most fundamental approach for achieving higher accuracy.

4. The Wireless Sensor Network (WSN) is spatially distributed autonomous sensor to sense special task, like ZigBee network forming simple interconnecting, low power, and low processing capability wireless devices.

英文科技论文写作

英文科技论文写作是进行国际学术交流必需的技能。一般而言，发表在专业英语期刊上的科技论文在文章结构和文字表达上都有其特定的格式和规定，只有严格遵循国际标准和相应刊物的规定，才能提高所投稿件的录用率。

撰写英文科技论文的第一步就是推敲结构。最简单有效的方法即采用IMRaD（导言、材料、结果及讨论）形式（Introduction，Materials and Methods，Results，and Discussion），这是英文科技论文最通用的一种结构方式。

IMRaD结构的逻辑体现在它能依次回答以下问题。

Introduction（引言）：研究的是什么问题？

Materials and Methods（材料和方法）：怎样研究这个问题？

Results（结果）：发现了什么？

Discussion（讨论）：这些发现意味着什么？

按照这个结构整体规划论文时，有一个方法值得借鉴，即剑桥大学爱席比教授提出的"概念图"。首先在一张大纸上（A3或A4纸，横放）写下文章题目（事先定好题目很重要），然后根据IMRaD的结构确定基本的段落主题，把他们写在不同的方框内。你可以记录任何你脑海中闪现的可以包括在该部分的内容，诸如段落标题、图表以及需要进一步阐述的观点等等，把它们写在方框附近的圈内，并用箭头标示它们所属的方框。画概念图的阶段也是自由思考的阶段，在此过程中不必拘泥于细节。哪些东西需要纳入文章？还需要做哪些工作？是找到某文献的原文，还是补画一张图表，或者需要再查找某个参考文献？当你发现自己需要再加进一个段落时就在概念图中添加一个新框。如果你发现原来的顺序需作调整，那就用箭头标示新的顺序。绘制概念图的过程看似如儿童游戏，但其意义重大，它可以给你自由思考的空间，并通过图示的方式记录你思维发展的过程。这便是写论文的第一步：从整体考虑文章结构，思考各种组织文章的方法，准备好所需的资料，随时记录出现的新想法。采用这个方法，不论正式下笔时是从哪一部分写起，都能够能做到大局不乱。

英文科技论文的基本格式包括：

Title——论文题目

Author(s)——作者姓名

Affiliation(s) and address(es)——联系方式

Abstract——摘要

Keywords——关键词

Body——正文

Acknowledgements——致谢，可空缺

References——参考文献

Appendix——附录，可空缺

Resume——作者简介，视刊物而定

其中正文为论文的主体部分，分为若干章节。一篇完整的科技论文的正文部分由以下内容构成：

Introduction——引言/概述

Materials and Methods——材料和方法

Results——结果

Discussion——讨论

Conclusions——结论/总结

下面对科技论文主要构成部分的写法和注意事项进行详细介绍。

1. Title（论文题目）

由于只有少数人研读整篇论文，多数人只是浏览原始杂志或者文摘、索引的论文题目。因此须慎重选择题目中的每一个字，力求做到长短适中，概括性强，重点突出，一目了然。

论文题目一般由名词词组或名词短语构成，避免写成完整的陈述句。在必须使用动词的情况下，一般用分词或动名词形式。题目中介词、冠词小写，如果题目为直接问句，要加问号，间接问句则不用加问号。

具体写作要求如下。

（1）题目要准确地反映论文的内容。作为论文的"标签"，题目既不能过于空泛和一般化，也不宜过于繁琐，使人得不出鲜明的印象。为确保题目的含义准确，应尽量避免使用非定量的、含义不明的词，如"rapid""new"等；并力求用词具有专指性，如"a vanadium-iron alloy"明显优于"a magnetic alloy"。

（2）题目用语需简练、明了，以最少的文字概括尽可能多的内容。题目最好不超过10至12个单词，或100个英文字符（含空格和标点），如若能用一行文字表达，就尽量不要用两行（超过两行有可能会削弱读者的印象）。在内容层次很多的情况下，如果难以简短化，最好采用主、副题名相结合的方法，主副题名之间用冒号（:）隔开，如，Importance of replication in microarray gene expression studies: statistical methods and evidence from repetitive CDNA hybridizations (Proc Natl Acad Sci USA, 2000, 97(18): 9834—9839)，其中的副题名起补充、阐明作用，可起到很好的效果。

（3）题目要清晰地反映文章的具体内容和特色，明确表明研究工作的独到之处，力求简洁有效、重点突出。为使表达直接、清楚，以便引起读者的注意，应尽可能地将表达核心内容的主题词放在题名开头。如The effectiveness of vaccination against in healthy, working adults (N Engl J Med, 1995, 333: 889-893)中，如果作者用关键词vaccination作为题名的开头，读者可能会误认为这是一篇方法性文章：How to vaccinate this population? 相反，用effectiveness作为题名中第一个主题词，就直接指明了研究问题：Is vaccination in this population effective? 题名中应慎重使用缩略语。尤其对于可有多个解释的缩略语，应严加限制，必要时应在括号中注明全称。对那些全称较长，缩写后已得到科技界公认的，才可使用。为方便二次检索，题名中应避免使用化学式、上下角标、特殊符号（数字符号、希腊字母等）、公式、不常用的专业术语和非英语词汇（包括拉丁语）等。

（4）由于题目比句子简短，并且无需主、谓、宾，因此词序就变得尤为重要。特别是如果词语间的修饰关系使用不当，就会影响读者正确理解题目的真实含意。例如：Isolation of antigens from monkeys using complement-fixation techniques，可使人误解为"猴子使用了补体结合技术"。应改为：Using complement-fixation techniques in isolation of antigens from monkeys，即"用补体结合技术从猴体分离抗体"。

2. Author(s)（作者姓名）

按照欧美国家的习惯，名字（first name）在前，姓氏（surname / family name / last name）在后。

但我国人名地名标准规定，中国人名拼写均改用汉语拼音字母拼写，姓在前名在后。因此，若刊物无特殊要求，则应按我国标准执行。如果论文由几个人撰写，则应逐一写出各自的姓名。作者与作者之间用空格或逗号隔开。例如：Wan Da, Ma Jun。

3. Affiliation(s) and address(es)（联系方式）

摘要不宜太详尽，也不宜太简短，应将论文的研究体系、主要方法、重要发现、主要结论等在作者姓名的下方还应注明作者的工作单位、邮政编码、电子邮件地址或联系电话等。要求准确清楚，使读者能按所列信息顺利地与作者联系。例如：

Wei Min

Key Laboratory of Industrial Internet of Things & Networked Control, Ministry of Education

Chongqing University of Posts and Telecommunications (CQUPT)

2 Chongwen Road, Chongqing, 400065, China

Email: weimin@cqupt.edu.cn

也有刊物在论文标题页的页脚标出以上细节，在论文最后附上作者简介和照片。

4. Abstract（摘要）

摘要也称为内容提要，是对论文的内容不加注释和评论的简短陈述。其作用主要是为读者阅读、信息检索提供方便。

（1）摘要的构成要素

研究目的——准确描述该研究的目的，说明提出问题的缘由，表明研究的范围和重要性。

研究方法——简要说明研究课题的基本设计，结论是如何得到的。

结果——简要列出该研究的主要结果，有什么新发现，说明其价值和局限。叙述要具体、准确并给出结果的置信值。

结论——简要地说明经验，论证取得的正确观点及理论价值或应用价值，是否还有与此有关的其他问题有待进一步研究，是否可推广应用等。

（2）摘要的基本类型

摘要主要有两大类：资料性摘要（informative abstract），说明性摘要（descriptive abstract），还有一种为二者的结合，称为结合型。一般刊物论文所附摘要都属于这两类。另有结构型的摘要，遵循一定的格式和套路，便于计算机检索。

说明性摘要——只向读者指出论文的主要议题是什么，不涉及具体的研究方法和结果。它一般适用于综述性文章，也用于讨论、评论性文章，尤以介绍某学科近期发展动态的论文居多。

资料性摘要——适用于专题研究论文和实验报告型论文，它应该尽量完整和准确地体现原文的具体内容，特别强调指出研究的方法和结果、结论等。这类摘要大体按介绍背景、实验方法和过程、结果与讨论的格式写。

结合型摘要——是以上两种摘要的综合，其特点是对原文需突出强调的部分做出具体的叙述，对于较复杂，无法三言两语概括的部分则采用一般性的描述。

结构性摘要——随着信息科学和电子出版物的发展，近年来又出现了一种新的摘要形式即结构性摘要。这类摘要先用短语归纳要点，再用句子加以简明扼要的说明，便于模仿和套用，能规范具体地将内容表达出来，方便审稿，便于计算机检索。

（3）摘要的撰写要求

（a）确保客观而充分地表述论文的内容，适当强调研究中的创新、重要之处（但不要使用评价性语言）；尽量包括论文中的主要论点和重要细节（重要的论证或数据）。

（b）要求结构严谨、语义确切、表述简明、一般不分段落；表述要注意逻辑性，尽量使用指示性的词语来表达论文的不同部分（层次），如使用"We found that..."表示结果；使用"We suggest that..."表示讨论结果的含义等。

（c）排除在本学科领域方面已成为常识的或科普知识的内容；尽量避免引用文献，若无法回避使用引文，应在引文出现的位置将引文的书目信息标注在方括号内；不使用非本专业的读者尚难于清楚理解的缩略语、简称、代号，如确有需要（如避免多次重复较长的术语）使用非同行熟知的缩写，应在缩写符号第一次出现时给出其全称；不使用一次文献中列出的章节号、图号、表号、公式号以及参考文献号。

（d）要求使用法定计量单位以及正确地书写规范字和标点符号；众所周知的国家、机构、专用术语尽可能用简称或缩写；为方便检索系统转录，应尽量避免使用图、表、化学结构式、数学表达式、角标和希腊文等特殊符号。

（e）摘要的长度：应符合相应刊物规定，大多数实验研究性文章，字数在1000~5000字的，其摘要长度限于100~250个英文单词。

（f）摘要的时态：摘要所采用的时态视情况而定，应力求表达自然、妥当。写作中可大致遵循以下原则：①介绍背景资料时，如果句子的内容是不受时间影响的普遍事实，应使用现在式；如果句子的内容为对某种研究趋势的概述，则使用现在完成式。②在叙述研究目的或主要研究活动时，如果采用"论文导向"，多使用现在式（如This paper presents...）；如果采用"研究导向"，则使用过去式（如This study investigated...）。③概述实验程序、方法和主要结果时，通常用现在式，如We describe a new molecular approach to analyzing ...④叙述结论或建议时，可使用现在式、臆测动词或may, should, could等助动词，如We suggest that climate instability in the early part of the last interglacial may have...

（g）摘要的人称和语态：作为一种可阅读和检索的独立使用的文体，摘要一般只用第三人称而不用其他人称来写。有的摘要出现了"我们""作者"作为陈述的主语，这会减弱摘要表述的客观性，有时也会导致逻辑上讲不通。由于主动语态的表达更为准确，且更易阅读，因而目前大多数期刊都提倡使用主动语态，国际知名科技期刊"Nature""Cell"等尤其如此。

5. Keywords（关键词）

关键词是为了满足文献标引或检索工作的需要而从论文中取出的词或词组。国际标准和我国标准均要求论文摘要后标引3~8个关键词。关键词既可以作为文献检索或分类的标识，又是论文主题的浓缩。读者从中可以判断论文的主题、研究方向与方法等。关键词包括主题词和自由词两类：主题词是专门为文献的标引或检索而从自然语言的主要词汇中挑选出来的，并加以规范化了的词或词组；自由词则是未规范的即还未收入主题词表中的词或词组。关键词以名词或名词短语居多，如果使用缩略词，则应为公认和普遍使用的缩略语，如IP、CAD、CPU，否则应写出全称，其后用括号标出其缩略语形式。

Unit 7

Mobile Communication Technology

Passage A   3G & 4G Technology

Passage B   Modulation (FSK, GFSK)

Passage C   5G and LTE-M

Passage A 3G & 4G Technology

I Overview

3G, short form of third generation, is the third generation of mobile telecommunications technology. This is based on a set of standards used for mobile devices and mobile telecommunications use services and networks that comply with the International Mobile Telecommunications-2000 (IMT-2000) specifications by the International Telecommunication Union. 3G finds application in wireless voice telephony, mobile Internet access, fixed wireless Internet access, video calls and mobile TV.

TD-SCDMA is the first TDD standard to be deployed in wide area mobile networks on a large scale. China Mobile commercially launched TD-SCDMA 3G in 2009. Statistics show that as of April 2014 here were already 230 million TD-SCDMA users. 3G cellular networks have been widely deployed and the large coverage of 3G networks allows 3G users to download files easily from the Internet with modest latency.

4G, short for fourth generation, is the fourth generation of mobile telecommunications technology, succeeding 3G. A 4G system must provide capabilities defined by ITU in IMT Advanced. Potential and current applications include amended mobile web access, IP telephony, gaming services, high-definition mobile TV, video conferencing, 3D television, and cloud computing. TD-LTE/ TD-LTE-Advanced are the long-term evolution of TD-SCDMA.

Theoretical rates of 100 Mbit/s for high mobility and up to 1 Gbit/s for low mobility are supposed to be reached with this generation. The main standards are LTE-Advanced and WiMAX. Their long-term goal would be to converge to form the 4G standard IMT-Advanced. The users of these networks are increasingly demanding in terms of throughput and reliability.

II. Development

Several telecommunications companies market wireless mobile Internet services as 3G, indicating that the advertised service is provided over a 3G wireless network. Services advertised as 3G are required to meet IMT-2000 technical standards, including standards for reliability and speed (data transfer rates). To meet the IMT-2000 standards, a system is required to provide peak data rates of at least 200 kbit/s (about 0.2 Mbit/s).

3G technology is the result of research and development work carried out by the International Telecommunication Union (ITU) in the early 1980s. 3G specifications and standards were developed in fifteen years. The technical specifications were made available to the public under the name IMT-2000. The communication spectrum between 400 MHz to 3 GHz was allocated for 3G. Both the government and communication companies approved the 3G standard. The first pre-commercial 3G network was launched by NTT DoCoMo in Japan in 1998, branded as FOMA. It was first available in May 2001 as a pre-release (test) of W-CDMA technology. The first commercial launch of 3G was also by NTT DoCoMo in Japan on 1 October 2001, although it was initially somewhat limited in scope; broader availability of the system was delayed by apparent concerns over its reliability.

However, 3G service cannot seamlessly integrate the existing wireless technologies (e.g., GSM, wireless LAN, and Bluetooth), and cannot satisfy users' fast growing needs for high data rates. Thus, most major cellular operators worldwide plan to deploy fourth generation (4G) networks to provide much higher data rates(up to hundreds of megabits per second) and integrate heterogeneous wireless technologies.

In March 2008, the International Telecommunications Union-Radio communications sector(ITU-R) specified a set of requirements for 4G standards, named the International Mobile Telecommunications Advanced (IMT-Advanced) specification, setting peak speed requirements for 4G service at 100 megabits per second (Mbit/s) for high mobility communication (such as from trains and cars) and 1 gigabit per second (Gbit/s) for low mobility communication (such as pedestrians and stationary users).

4G LTE service providers in the US, AT&T, TMobile and Verizon operate using both lower and upper bands. LTE supports both the versions of duplexing (i.e.,transmit and receive) methods Time-Division Duplexing(TDD), and Frequency-Division Duplexing (FDD) combined with the downlink modulation scheme, Orthogonal Frequency-Division Multiple Access (OFDMA) to achieve maximum peak downlink data rate of 100 Mbps.

LTE is currently being deployed commercially by mobile operators around the world. This advanced wireless core network technology helps mobile operators to build an overlay network over the existing cellular radio networks for supporting high speed data bit rate and other advanced value-added application services.

The 4G system was originally envisioned by the Defense Advanced Research Projects Agency (DARPA). The DARPA selected the distributed architecture and end-to-end Internet protocol (IP), and believed at an early stage in peer-to-peer networking in which every mobile device would be both a transceiver and a router for other devices in the network, eliminating the spoke-and-hub weakness of 2G and 3G cellular systems. Since the 2.5G GPRS system, cellular systems have provided dual infrastructures: packet switched nodes for data services, and circuit switched nodes for voice calls. In 4G systems, the circuit-switched infrastructure is abandoned and only a packet-switched network is provided, while 2.5G and 3G systems require both packet-switched and circuit-switched network nodes, i.e. two infrastructures in parallel. This means that in 4G, traditional voice calls are replaced by IP telephony.

Ⅲ. New Words

reliability [rɪˈlaɪəˈbɪlətɪ]	n.	可靠性
specification [ˌspesɪfɪˈkeɪʃ(ə)n]	n.	规格；说明书；详述
launch [lɔːntʃ]	vt.	启动；发布
scope [skəʊp]	n.	范围；导弹射程
integrate [ˈɪntɪgreɪt]	vt.	使……完整；求……的积分
heterogeneous [ˌhetərəˈdʒiːnɪəs]	adj.	多相的；异种的
bluetooth [ˈbluːtuːθ]	n.	蓝牙技术
duplexing [ˈdjuːpleksɪŋ]	n.	双工，复式；双炼法

Ⅳ. Phrases

International Mobile Telecommunications	国际移动通信
cellular radio networks	蜂窝移动网络

Ⅴ. Abbreviations

TD-SCDMA　Time Division-Synchronization Code Division Multiple Access
　　　　　　　　　　　　　　　　　　　　时分同步的码分多址技术
TDD　Time Division Duplex　　　　　　　时分双工
ITU　International Telecommunication Union　国际电信联盟
GSM　Global System for Mobile Communication　全球移动通信系统
TDD　Time-Division Duplexing　　　　　时分双工
FDD　Frequency-Division Duplexing　　　频分双工
OFDMA　Orthogonal Frequency Division Multiple Access　正交频分多址接入
DARPA　Defense Advanced Research Projects Agency　美国国防部高级研究计划局

Passage B Modulation (FSK, GFSK)

Modulation is a process of encoding information from a message source in a manner suitable for transmission. It involves translating a baseband message signal to a pass band signal. The baseband signal is called the modulating signal and the pass band signal is called the modulated signal. Modulation can be done by varying certain characteristics of carrier waves according to the message signal. Demodulation is the reciprocal process of modulation which involves extraction of original baseband signal from the modulated pass band signal.

I. Trend

The move to digital modulation provides more information capacity, compatibility with digital data services, higher data security, better quality communications, and quicker system availability. Developers of communications systems face these constraints:

- available bandwidth
- permissible power
- inherent noise level of the system

Another layer of complexity in many new systems is multiplexing. Two principal types of multiplexing (or "multiple access") are TDMA (Time Division Multiple Access) and CDMA (Code Division Multiple Access). These are two different ways to add diversity to signals allowing different signals to be separated from one another.

Over the past few years a major transition has occurred from simple analog Amplitude Modulation (AM) and Frequency/Phase Modulation (FM/PM) to new digital modulation techniques. Trends of modulation is shown as Figure 7.1.

Examples of digital modulation include:

- QPSK (Quadrature Phase Shift Keying);
- FSK (Frequency Shift Keying);
- MSK (Minimum Shift Keying);
- QAM (Quadrature Amplitude Modulation);

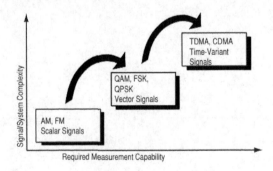

Figure 7.1 Trends of modulation

II. Influence character

1. Bit rate and symbol rate

To understand and compare different modulation format efficiencies, it is important to first understand the difference between bit rate and symbol rate. The signal bandwidth for the communications channel needed depends on the symbol rate, not on the bit rate.

Bit rate is the frequency of a system bit stream. The symbol rate is the bit rate divided by the number of bits that can be transmitted with each symbol. Symbol rate is sometimes called baud rate. Note that baud rate is not the same as bit rate. These terms are often confused. If more bits can be sent with each symbol, then the same amount of data can be sent in a narrower spectrum. This is why modulation formats that are more complex and use a higher number of states can send the same information over a narrower piece of the RF spectrum.

2. Spectrum (bandwidth) requirements

An example of how symbol rate influences spectrum requirements can be seen in eight-state Phase Shift Keying (8PSK). It is a variation of PSK. There are eight possible states that the signal can transition to at any time. The phase of the signal can take any of eight values at any symbol time. Since $2^3 = 8$, there are three bits per symbol. This means the symbol rate is one third of the bit rate. This is relatively easy to decode.

3. Symbol clock

The symbol clock represents the frequency and exact timing of the transmission of the individual symbols. At the symbol clock transitions, the transmitted carrier is at the correct I/Q (or magnitude/phase) value to represent a specific symbol (a specific point in the constellation).

III. Technology

1. FM

Frequency modulation (FM) involves changing of the frequency of the carrier signal according to message signal. As the information in frequency modulation is in the frequency of modulated signal, it is a nonlinear modulation technique. In this method, the amplitude of the carrier wave is kept constant (this is why FM is called constant envelope). FM is thus part of a more general class of modulation known as angle modulation.

FM systems require a wider frequency band in the transmitting media (generally several times as large as that needed for AM) in order to obtain the advantages of reduced noise and capture effect. FM transmitter and receiver equipment are also more complex than that used by amplitude modulation systems.

2. FSK

The binary version of FM is called Frequency Shift Keying or FSK. Here the frequency does not keep changing gradually over symbol time but changes in discrete amounts in response to a message similar to binary PSK where phase change is discrete.

FSK (Frequency Shift Keying) is used in many applications including cordless and paging systems. Some of the cordless systems include DECT(Digital Enhanced Cordless Telephone) and CT2(Cordless Telephone 2).

In FSK, the frequency of the carrier is changed as a function of the modulating signal (data) being transmitted. Amplitude remains unchanged. In binary FSK (BFSK or 2FSK), a "1" is represented by one frequency and a "0" is represented by another frequency.

3. MSK

MSK has a narrower spectrum than wider deviation forms of FSK. The width of the spectrum is also influenced by the waveforms causing the frequency shift. If those waveforms have fast transitions or a high slew rate, then the spectrum of the transmitter will be broad. In practice, the waveforms are filtered with a Gaussian filter, resulting in a narrow spectrum. In addition, the Gaussian filter has no time-domain overshoot, which would broaden the spectrum by increasing the peak deviation. MSK with a Gaussian filter is termed GMSK (Gaussian MSK).

4. BPSK

In binary phase shift keying (BPSK), the phase of a constant amplitude carrier signal is switched between two values according to the two possible signals m1 and m2 corresponding to binary 1 and 0, respectively. Normally, the two phases are separated by 180°. If the sinusoidal carrier has an amplitude A, and energy per bit then the transmitted BPSK signal is

$$S_{\text{BPSK}}(t) = m(t)\sqrt{\frac{2E_b}{T_b}}\cos(2\pi f_c t + \theta_c)$$

A typical BPSK signal constellation diagram is shown in Figure 7.2.

The probability of bit error for many modulation schemes in an AWGN channel is found using the Q-function of the distance between the signal points. In case of BPSK,

$$P_{\text{eBPSK}}(t) = Q(\sqrt{\frac{2E_b}{N_o}})$$

5. QPSK

The Quadrature Phase Shift Keying (QPSK) is a 4-ary PSK signal. The phase of the carrier in the QPSK takes 1 of 4 equally spaced shifts. Although QPSK can be viewed as a quaternary modulation, it is easier to see it as two independently modulated quadrature carriers. With this interpretation, the even (or odd) bits are used to modulate the in-phase component of the carrier, while the odd (or even) bits are used to modulate the quadrature-phase component of the carrier.

The QPSK transmitted signal (as shown in Figure 7.3) is defined by:

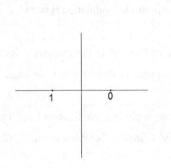

Figure 7.2 BPSK signal constellation

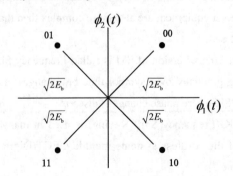

Figure 7.3 QPSK signal constellation

6. QAM

Quadrature Amplitude Modulation (QAM) is used in applications including microwave digital radio, DVB-C (Digital Video Broadcasting—Cable), and modems.

7. TDMA

Time-division multiplexing involves separating the transmitters in time so that they can share the same frequency. The simplest type is Time Division Duplex (TDD). This multiplexes the transmitter and receiver on the same frequency. TDMA (Time Division Multiple Access) multiplexes several transmitters or receivers on the same frequency. TDMA is used in the GSM digital cellular system and also in the USNADC-TDMA system.

The TDMA version of the North American Digital Cellular (NADC) system, achieves a 48 Kbits-per second data rate over a 30 kHz bandwidth or 1.6 bits per second per Hz. It is a π/4 DQPSK based system and transmits two bits per symbol. The theoretical efficiency would be two bits per second per Hz and in practice it is 1.6 bits per second per Hz.

Ⅳ. New Words

modulation [ˌmɒdjʊ'leɪʃən]	n.	调制；调整
encode [ɪn'kod]	vt.	编码，译码
transmission [træns'mɪʃən]	n.	传动装置，变速器；传递；传送；播送
complexity [kəm'pleksətɪ]	n.	复杂，复杂性；复杂错综的事物
bandwidth ['bændwɪdθ]	n.	带宽；[通信] 频带宽度
spectrum ['spektrəm]	n.	光谱；频谱；范围；余象

Ⅴ. Phrases

available bandwidth	可用带宽
permissible power	允许功耗

Ⅵ. Abbreviations

TDMA	Time Division Multiple Access	时分多址
CDMA	Code Division Multiple Access	码分多址
AM	Amplitude Modulation	波幅调制
FM/PM	Frequency/Phase Modulation	频率、相位调制
QPSK	Quadrature Phase Shift Keying	正交相移键控
FSK	Frequency Shift Keying	频移键控
MSK	Minimum Shift Keying	最小频移键控
QAM	Quadrature Amplitude Modulation	正交调幅
NADC	North American Digital Cellular	北美数字蜂窝

Passage C 5G and LTE-M

Ⅰ. 5G

5G is the next generation mobile and wireless connectivity system. Key requirements are that it should offer far greater capacity and be more responsive to users' needs, more energy-efficient and more cost-effective than anything that has gone before.

The need for the 5G network is being driven by the ever-increasing demand for mobile data and the emergence of the IoT, through which billions of devices will become connected. In the future, technologies such as Smart Cities, non-intrusive healthcare and advanced logistics will require much shorter network response times to enable very rapid reactions. At the same time there is a pressing need to reduce end user costs (to ensure applications are widely accessible) and to minimise energy consumption.

To meet these requirements, a range of tactics need to be employed including: more accurately predicting user demand so that applications perform bandwidth-heavy tasks making better use of all available wireless networks for ultra-high reliability.

Ultimately, the 5G network will need to be flexible enough to evolve, adapt and grow for as yet unforeseeable applications – just as the internet has.

1. Three Very Distinct 5G Network visions

If 5G appears and reflects these prognoses, then the major difference, from a user point of view, between 4G and 5G must be something other than faster speed (increased peak bit rate). For example, higher number of simultaneously connected devices, higher system spectral efficiency (data volume per area unit), lower battery consumption, lower outage probability (better coverage), high bit rates in larger portions of the coverage area, lower latencies, higher number of supported devices, lower infrastructure deployment costs, higher versatility and scalability, or higher reliability of communication.

GSMHistory.com has recorded three very distinct 5G network visions that had emerged by 2014:

A super-efficient mobile network that delivers a better performing network for lower investment cost. It addresses the mobile network operators' pressing need to see the unit cost of data transport falling at roughly the same rate as the volume of data demand is rising. It would be a leap forward in efficiency based on the IET Demand Attentive Network (DAN) philosophy.

A super-fast mobile network comprising the next generation of small cells densely clustered to give a contiguous coverage over at least urban areas and getting the world to the final frontier of true "wide-area mobility." It would require access to spectrum under 4 GHz perhaps via the world's first global implementation of Dynamic Spectrum Access.

A converged fiber-wireless network that uses, for the first time for wireless Internet access, the millimeter wave bands (20 – 60 GHz) so as to allow very-wide-bandwidth radio channels able to support data-access speeds of up to 10 Gbit/s. The connection essentially comprises "short" wireless links on the end of local fiber optic cable. It would be more a "nomadic" service (like Wi-Fi) rather than a wide-area "mobile" service.

2. Research & Development Projects

In 2008, the Republic of Korea IT R&D program of "5G mobile communication systems based on beam-division multiple access and relays with group cooperation" was formed.

In 2012, the UK Government announced the establishment of a 5G Innovation Centre at the University of Surrey – the world's first research center set up specifically for 5G mobile research.

In 2012, NYU WIRELESS was established as a multidisciplinary research center, with a focus on 5G wireless research, as well as its use in the medical and computer-science fields. The center is funded by the National Science Foundation and a board of 10 major wireless companies (as of July 2014) that serve on the Industrial Affiliates board of the center. NYU WIRELESS has conducted and published channel measurements that show that millimeter wave frequencies will be viable for multigigabit-per-second data rates for future 5G networks.

In 2012, the European Commission, under the lead of Neelie Kroes, committed 50 million euros for research to deliver 5G mobile technology by 2020. In particular, The METIS 2020 Project is driven by several telecommunication companies, and aims at reaching world-wide consensus on the future global mobile and wireless communication system. The METIS overall technical goal is to provide a system concept that supports 1,000 times higher mobile system spectral efficiency, compared to current LTE deployments. In addition, in 2013, another project has started, called 5GrEEn, linked to project METIS and focusing on the design of green 5G mobile networks. Here the goal is to develop guidelines for the definition of a new-generation network with particular emphasis on energy efficiency, sustainability and affordability.

In November 2012, a research project funded by the European Union under the ICT Programme FP7 was launched under the coordination of IMDEA Networks Institute (Madrid, Spain): iJOIN (Interworking and Joint Design of an Open Access and Backhaul Network Architecture for Small Cells based on Cloud Networks). iJOIN introduces the novel concept of the radio access network (RAN) as a service (RANaaS), where RAN functionality is flexibly centralized through an open IT platform based on a cloud infrastructure. iJOIN aims for a joint design and optimization of access and backhaul, operation and management algorithms, and architectural elements, integrating small cells, heterogeneous backhaul and centralized processing. Additionally to the development of technology candidates across PHY, MAC, and the network layer, iJOIN will study the requirements, constraints and implications for existing mobile networks, specifically 3GPP LTE-A.

In January 2013, a new EU project named CROWD (Connectivity management for eneRgy Optimised Wireless Dense networks) was launched under the technical supervision of IMDEA Networks Institute, to design sustainable networking and software solutions for the deployment of very dense, heterogeneous wireless networks. The project targets sustainability targeted in terms of cost effectiveness and energy efficiency. Very high density means 1000x higher than current density (users per square meter). Heterogeneity involves multiple dimensions, from coverage radius to technologies (4G/LTE vs. Wi-Fi), to deployments (planned vs. unplanned distribution of radio base stations and hot spots).

In September 2013, the Cyber-Physical System (CPS) Lab at Rutgers University, NJ, started to work on dynamic provisioning and allocation under the emerging cloud radio-access network (C-RAN). They have shown that the dynamic demand-aware provisioning in the cloud will decrease the energy consumption while increasing the resource utilization.They also have implemented a test bed for feasibility of C-RAN and developed new cloud-based techniques for interference cancellation. Their project is funded by the National Science Foundation.

In November 2013, Chinese telecom equipment vendor Huawei said it would invest $600 million in research for 5G technologies in the next five years.The company's 5G research initiative does not include investment to productize 5G technologies for global telecom operators. Huawei will be testing 5G technology in Malta.

In 2015, Huawei and Ericsson were testing 5G-related technologies in rural areas in northern Netherland.

In July 2015, the 5GNORMA project was launched. The key objective of 5G NORMA is to develop a conceptually novel, adaptive and future-proof 5G mobile network architecture. The architecture is enabling unprecedented levels of network customisability, ensuring stringent performance, security, cost and energy requirements to be met; as well as providing an API-driven architectural openness, fuelling economic growth through over-the-top innovation. With 5G NORMA, leading players in the mobile ecosystem aim to underpin Europe's leadership position in 5G.

In July 2015, the European research project mmMAGIC was launched. The mmMAGIC project will develop new concepts for mobile radio access technology (RAT) for mmwave band deployment. This is a key component in the 5G multi-RAT ecosystem and will be used as a foundation for global standardization. The project will enable ultrafast mobile broadband services for mobile users, supporting UHD/3D streaming, immersive applications and ultra-responsive cloud services. A new radio interface, including novel network management functions and architecture components will be designed taking as guidance 5G PPP's KPI and exploiting the use of novel adaptive and cooperative beam-forming and tracking techniques to address the specific challenges of mm-wave mobile propagation. The ambition of the project is to pave the way for a European head start in 5G standards and to strengthen European competitiveness. The consortium brings together major infrastructure vendors, major European operators, leading research institutes and universities, measurement equipment vendors and one SME.

In July 2015, IMDEA Networks launched the Xhaul project, as part of the European H2020 5G Public-Private Partnership (5G PPP). Xhaul will develop an adaptive, sharable, cost-efficient 5G transport network solution integrating the fronthaul and backhaul segments of the network. This transport network will flexibly interconnect distributed 5G radio access and core network functions, hosted on in-network cloud nodes. Xhaul will greatly simplify network operations despite growing technological diversity. It will hence enable system-wide optimisation of Quality of Service (QoS) and energy usage as well as network-aware application development. The Xhaul consortium comprises 21 partners including leading telecom industry vendors, operators, IT companies, small and medium-sized enterprises and academic institutions.

In July 2015, the European 5G research project Flex5Gware was launched. The objective of Flex5Gware is to deliver highly reconfigurable hardware (HW) platforms together with HW-agnostic software (SW) platforms targeting both network elements and devices and taking into account increased capacity, reduced energy footprint, as well as scalability and modularity, to enable a smooth transition from 4G mobile wireless systems to 5G. This will enable that 5G HW/SW platforms can meet the requirements imposed by the anticipated exponential growth in mobile data traffic (1000 fold increase) together with the large diversity of applications (from low bit-rate/power for M2M to interactive and high resolution applications).

In July 2015, the SUPERFLUIDITY project, part of the European H2020 Public-Private Partnership (5G PPP) and led by CNIT, an Italian inter-university consortium, was started. The SUPERFLUIDITY consortium comprises telcos and IT players for a total of 18 partners. In physics, superfluidity is a state in which matter behaves like a fluid with zero viscosity. The SUPERFLUIDITY project aims at achieving superfluidity in the Internet: the ability to instantiate services on-the-fly, run them anywhere in the network (core, aggregation, edge) and shift them transparently to different locations. The project tackles crucial shortcomings in today's networks: long provisioning times, with wasteful over-provisioning used to meet variable demand; reliance on rigid and cost-ineffective hardware devices; daunting complexity emerging from three forms of heterogeneity: heterogeneous traffic and sources; heterogeneous services and needs; and heterogeneous access technologies, with multi-vendor network components. SUPERFLUIDITY will provide a converged cloud-based 5G concept that will enable innovative use cases in the mobile edge, empower new business models, and reduce investment and operational costs.

On January 29, 2016, Google revealed that they were developing a 5G network called SkyBender. They planned to distribute this connection through sun-powered drones.

In mid-March 2016, the UK government confirmed plans to make the UK a world leader in 5G. Plans for 5G are little more than a footnote in the country's 2016 budget, but it seems the UK government wants it to be a big focus going forward.

II. LTE-M

The Internet of Things(IoT) refers to interconnection and exchange of data among devices. To support IoT, Machine-to-Machine(M2M) communication is needed. M2M is defined as data communication among devices without the need of human interaction.

Currently, M2M services are supported using cellular systems such as GSM. With the widespread introduction of LTE and decommissioning of legacy systems, migration of M2M devices to LTE is under consideration by many cellular operators. In LTE Rel-12, low cost M2M devices with material cost comparable to EGPRS devices are being introduced.

III. Narrowband LTE-M Design

For M2M communication, a narrowband system is attractive due to the following reasons—

- Low cost especially on the device side. Narrow bandwidth requires less expensive RF components. In addition, there is also a cost reduction on the baseband side due to the lower data rates.
- Coverage improvement due to the ability to concentrate transmission power in narrow bandwidth.
- Efficient spectrum utilization as smaller bandwidth is needed. For example, LTE-M can be deployed by refarming only one GSM channel, or it can be deployed on a guard band of an existing LTE deployment.

The motivation for reusing LTE design for narrowband M2M system is to take advantage of existing technology as well as installed system base. By making LTE-M compatible with LTE, it is possible to reuse the same hardware and also to share spectrum without coexistence issues. In addition, LTE-M can simply plug into the LTE core network. This allows all network services such as authentication, security, policy, tracking, and charging to be fully supported.

Figure 7.4 illustrates the design for LTE-M with 180 kHz occupied bandwidth using either existing or modified LTE channels. Some of the channels such as data channels already fit into 180 kHz (equivalent to 1 physical resource block, or PRB, in LTE). Other channels will need some modifications to fit into 180 kHz. However, the same design principle will be kept, allowing as much reuse from LTE as possible.

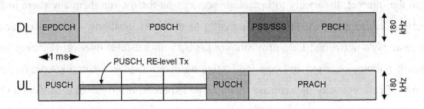

Figure 7.4　LTE-M design of 180kHz bandwidth

An important feature of LTE-M is that it shares the same numerology as LTE. This allows for sharing spectrum between the two systems without causing mutual interference. In the uplink, the two systems can be frequency-multiplexed together. In the downlink, however, LTE contains TDM control portion. In this case, the LTE-M downlink channels will need to be punctured as shown in Figure 7.5.

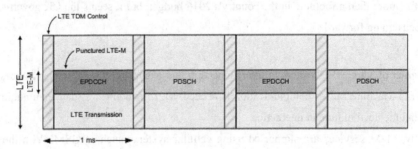

Figure 7.5　LTE-M multiplexed into LTE (DL).

Sharing of the same PRB between LTE-M and LTE in time allows for more efficient use of the spectrum and seamless increase in LTE-M capacity by multiplexing additional LTE-M channels into LTE PRBs as more M2M devices are added to the network. In addition, although they are two separate systems, they can be supported using the same ENB hardware.

Ⅳ. Analysis

In this section, coexistence of GSM and when GSM spectrum is re-farmed to deploy LTE-M, is analyzed. An LTE-M channel can easily be placed within a 200-kHz-wide GSM channel. We consider the case where the LTE-M channel replaces a GSM channel in the middle of the GSM spectrum, as depicted in Fig. 7.6(a). Here there is mutual interference between the LTE-M channel and the GSM channel on either side of it due to spectral leakage. To reduce this interference, a guard band can be used between the LTE-M channel and each of its two adjacent channels. Fig. 7.6(b) illustrates the re-farming of two GSM channels to deploy a single LTE-M channel, resulting in a 100-kHz guard band.

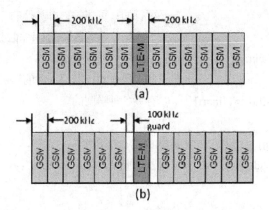

Figure 7.6   LTE-M and GSM channel allocation(a) without guard band and(b) with 100 kHz guard band.

The effects of adjacent channel can be characterized in terms of the adjacent leakage ratio(ACLR) and the adjacent channel selectivity(ACS), which together determined the adjacent channel interference ratio(ACIR), in the receiver of the victim system due to the aggressor system.

Coexistence performance is analysed for the downlink through system simulations. LTE-M and GSM channels in the 900 MHz frequency band are assumed. Thus, the interference scenarios that are considered for the study are:

- GSM BTS to LTE-M UE
- LTE-M ENB to GSM UE

This section shows that a narrowband LTE-M system can be deployed using one GSM channel and can also be multiplexed with LTE. Analysis shows that the system can coexist with a GSM system and requires minimal changes to the deployed LTE system. In addition, the LTE-M system offers significant capacity and coverage extension of 20 dB beyond LTE is feasible.

Ⅴ. New Words

interconnection [,ɪntəkə'nekʃən]	n. 互连；互相连络
interaction [ɪntər'ækʃ(ə)n]	n. 相互作用；相互影响
cellular ['seljʊlə]	n. 移动电话；单元
decommissioning [,di:kə'mɪʃənɪŋ]	n. 退役；停运
legacy ['legəsɪ]	n. 遗赠，遗产
migration [maɪ'greɪʃ(ə)n]	n. 迁移；移民；移动
narrowband ['nærəʊbænd]	n. 窄频带
sideband ['saɪdbænd]	n. 边（频）带，旁（频）带
spectrum ['spektrəm]	n. 光谱；频谱；范围；余象
hardware ['hɑːdweə]	n. 硬件
seamless ['siːmlɪs]	adj. 无缝的；无缝合线的；无伤痕的
coexistence [,kəʊɪg'zɪstəns]	n. 共存
aggressor [ə'gresə]	n. 侵略者
scenario [sɪ'nɑːrɪəʊ]	n. 情景

feasible ['fiːzɪb(ə)l]	*adj.* 可行的；可能的；可实行的
energy-efficient	*adj.* 节能的，节约能源的；高能效的
cost-effective [kɔːst ɪ'fektɪv]	*adj.* 划算的；成本效益好的
non-intrusive	*adj.* 非入侵式的；非侵入性的
multidisciplinary [ˌmʌltɪdɪsə'plɪnərɪ]	*adj.* 有关各种学问的
backhaul ['bækhɔːl]	*n.* 回程
integrating ['ɪntɪgreɪtɪŋ]	*n.* 集成化；综合化
superfluidity [suːpəfluː'ɪdɪtɪ]	*n.* [低温] 超流动性；超流态

VI. Phrases

narrow bandwidth	窄带宽
plug into	把（电器）插头插入；接通
fit into	（使）适合，适应；符合
coexistence of	共存情况；共存
DAN Demand Attentive Network	细致网络需求
Dynamic Spectrum Access	动态频谱接入

VII. Abbreviations

EGPRS Enhanced General Packet Radio Service	增强型通用分组无线业务
RF radio frequency	无线电频率
PRB Physical Resource Bearer	物理资源承载
TDM Time Division Multiplex	时分多路转换
ENB effective noise bandwidth	有效噪声带宽
ACLR Adjacent Channel Leakage Ratio	相邻频道泄漏比
ACS Adjacent Channel Selectivity	相邻通道选择
ACIR Adjacent Channel Interference Ratio	邻道干扰比
FP7 Framework Program 7	第七框架计划
RAN radio access network	无线接入网络
CROWD Connectivity management for eneRgy Optimised Wireless Dense networks	能源密度优化无线网络的连接管理
CPS Cyber-physical system	信息物理系统

Exercises

Translate the following sentences into Chinese or English.

1. There is some restriction in 1G & 2G technologies for speed & security so to improve the speed & security there is need of new technology & therefore the 3G comes in picture, representing the convergence of various 2G wireless telecommunications systems into a single global system that includes both terrestrial and satellite components.

2. CDMA (Code Division Multiple Access) which comes under 3G technology uses spread spectrum technology to break up speech into small, digitized segments and encodes them to identify each call.

3. 以3G和WLAN为代表的异构无线网络融合是下一代无线网络发展的必然趋势。

4. 移动互联网的发展带动了4G移动通信的加速发展，其快速发展对整个信息产业其他领域的影响很大。

参考译文 Passage A 3G&4G技术

Ⅰ. 概述

3G,是第三代移动通信技术的简称。基于移动设备和移动通信应用服务和网络的一整套标准,3G符合由国际电信联盟制定的国际移动通信-2000(IMT-2000)规格。3G有很多应用,如无线语音电话、移动互联网接入、固定无线上网、视频通话和手机电视。

TD-SCDMA是第一个TDD标准,将被大规模地部署于广域移动网络。2009年,中国移动商业化地发行了TD-SCDMA 3G。统计数据显示,截至2014年4月,TD-SCDM用户已有2.3亿。3G蜂窝网已广泛部署,其大范围的网络覆盖使3G用户易于从互联网上下载文件,并支持适度延迟。

4G,是继第三代之后的第四代移动通信技术的简称。4G系统必须提供由国际电信联盟(ITU)IMT-Advanced规定的功能。当前正在使用的和潜在的应用包括移动网络接入、IP电话、游戏服务、高清手机电视、视频会议、3D电视、云计算。TD-LTE和TD-LTE-Advanced是TD-SCDMA的长期演进。

这一代移动通信技术应该达到100 Mbit/s的高流动性和1 Gbit/s的低流动性。主要的标准是LTE-Advanced和WiMAX,其长期目标是汇聚以形成4G标准IMT-Advanced。这些网络的用户在吞吐量和可靠性方面的要求越来越高。

Ⅱ. 发展

一些电信公司部署无线移动互联网为3G服务,这说明发布的服务是由3G无线网络提供的。发布的3G服务必须符合IMT-2000 技术标准,包括可靠性和速度标准(数据传输率)。为了满足IMT-2000标准,一个系统需提供至少200 kbit/s(约0.2 Mbit/s)的峰值数据速率。

早在20世纪80年代,国际电信联盟(ITU)就部署相关研究和开发工作,研发出了3G技术。十五年来,3G的规格和标准也有所发展,公众可使用IMT-2000技术规格为3G网络分配400 MHz到3 G的通信频谱。政府和通信公司都批准了3G标准。日本NTT DoCoMo于1998年推出首个试用性的商用3G网络,并将其命名为FOMA。于2001年5月,作为预发布(测试)的W-CDMA技术首次可供使用。3G的第一次商业活动,也是由日本电报电话公司在2001年10月1日举办的,尽管最初在范围上有所限制;出于对其可靠性的担忧,系统广泛的被推迟了。

然而,3G服务不能无缝集成现有的无线技术(如全球移动通信系统、无线局域网和蓝牙),也不能满足用户对高数据速率快速增长的需求。因此,全球的很多移动运营商计划部署四代(4G)网络,以提供更高的数据速率(高达每秒数百兆)和集成异构的无线技术。

2008年3月,国际电信联盟无线电通信部门(ITU-R)规定了用于4G标准的要求,名为高级国际移动通信(IMT-Advanced)规范,设置4G服务的峰值速度要求,100Mbit/s用于高移动通信(例如,火车和汽车),1Gbit/s用于低移动通信(例如,行人和静态用户)

美国的4G LTE服务供应商AT&T、T移动和威瑞森同时利用低频和高频进行无线通信。LTE支持两种双工(也就是,传送和接收)版本,时分双工以及频分双工,通过与下行线调制方案,和正交频分多址接入相结合,以实现100Mbps的最大峰值下行线数据速率。

目前,移动运营商正在全世界对LTE进行商业部署。这种先进的无线核心网络技术,有助于移动运营商在现有的蜂窝无线网络基础上建立一个覆盖网络,支持高速数据比特率和其他先进的增值应用服务。

4G系统最初是由美国国防部高级研究计划局（DARPA）提出的。DARPA挑选分布式架构和端对端互联网协议，并相信在早期的对等网络中，每一个移动设备都同时是接收器和路由器，克服了2G和3G蜂窝系统中的中心与分支的弱点。从2.5G的GPRS系统以来，蜂窝系统已经提供了双重基础设施：关于数据服务的分组交换节点，以及关于语音的电路交换节点。在4G系统中，电路交换基础设施被舍弃，并且只提供分组交换网络，而2.5G和3G系统需要分组交换和电路交换网络节点，也就是说，两个平行的基础设施。这意味着在4G系统中，IP电话替代了传统的语音电话。

参考答案

1. 1G和2G技术在速度和安全性方面有一些限制，为提高速度和安全性，需要新的技术，从而导致了3G技术的出现，该技术将各种不同的2G无线通信系统融合成一套单一的全球系统，包括地面和卫星部分。

2. 码分多址（CDMA）在3G技术中，使用扩频技术将语音拆分成小的、数字化的片段，并进行编码，以识别每个电话。

3. Heterogeneous wireless network integration, typically 3G and WLAN integration, is an inevitable trend.

4. The development of mobile internet has promoted the rapid development of fourth-generation mobile communication, and the rapid development makes a great influence on the whole information industry in other areas.

如何用英语撰写科技论文

科技论文的写作就是用最平实的语言陈述一项发现或研究成果。英语科技论文的写作最关键的是具备用英语清楚表述的能力。英语是当前科技领域的国际语言，想要你的观点、研究让世界各地的科技同行获知，那么就要用英语来撰写论文。选词是科技论文中首要考虑的事项。

1. 用准确的词语

事实上，科技论文中大多数句子的问题不是语法，而是词语的选择不当。请看下面的例句：

The current remained **enhanced** for **several** hours.

这句话中，"enhanced"和"several"用在此处不合适。因为这两个词语不精确，有歧义，不能准确地表达作者所描述的事物。将"enhanced"改成"increased"，可表示增加，并能清楚表达作者的意思。为了达到准确清楚的标准，也要将"several"这个十分模糊的词语改成精确的数字。

2. 用简洁的词语

本来母语非英语的科技人士用英语表达就较为困难，所以更加需要避免使用复杂的英语词语。除此之外，科技论文不仅要求准确，同样要求简洁。请看下面的例句：

Fractions of 0.6ml were collected, **reduced to dryness**, and dissolved in 2.85% methanol(v/v) prior to being sequenced.

科技论文中有许多专业词汇，读起来比较繁重。因此，就应该选择简洁的非专业词汇。例句中的"reduced to dryness"改成"dried"，表达了相同的意思，同时使句子更加易读。

3. 不用不必要的词语和短语

一再表明，撰写科技论文最重要的原则是准确、简洁，即在清楚表述的前提下，尽量用词简洁，不用不必要的词语。请看下面的例句：

High PH values **have been observed to** occur in areas that **have been determined to** have few pine trees.

以上例句中黑体部分都是可以删除的，修改之后的句子简洁，明了。

常见的不必使用的短语：

It is(was) well known that…

In a previous study…

It is(was) demonstrated that…

It is (was) shown that…

4. 缩略语的选择

使用缩略语时要多加考虑，因为大量的缩略语会使读者困惑，所以要少使用。如果一个很长的词汇同页出现频率过高，以缩略语代替是合理的。但是，同一篇论文中缩略语的使用不应超过5个。此外，还应避免在一段话中同时使用多个缩略语。请看下面例句：

MPTP is converted by **MAOB** to **MPP**, which reaches **SNpc** nerve cells via DA uptake systems.

这句话对专业人员来说很好理解，但大多数读者却无法理解。因此，在写作中，对于必须使用的缩略语，应在第一次出现时进行定义，或脚注中标注出来。

5. 专业术语的选择

科技论文中，专业名称、专业术语较多，故撰写论文时正确使用和正确选择尤为重要。

常见的专业术语：

RFID　Radio-frequency identification　　　　　射频识别
UHFID　Ultra High Frequency Identification　　超高频识别
UHF　　Ultra High Frequency　　　　　　　　超高频
LF　　　Low Frequency　　　　　　　　　　　低频
HF　　　High Frequency　　　　　　　　　　　高频

6. 动词时态和介词的用法

母语为非英语的科技人士在撰写科技论文时，除了选词这一难点，英语作为第二语言的特殊语法问题也难倒了很多人，特别是有关动词时态和介词的用法。

科技论文中较为常用的是现在时态和过去时态。通常，表示实验方法、结果及特定结论部分应全部或大多数用过去时态。例如，描述实验结果时使用过去时态，论文的前言部分则使用现在时态，因为前言部分多为普遍认可的概念和观点，而已经发表的结果通常被视为"事实"。

论文的讨论部分通常要使用几种不同的时态。讨论部分是探讨你的研究与前人研究的关系，通常用到几种时态。例如，陈述已发表的研究成果时用现在时态，而总结该论文所陈述研究工作的内容时应该使用过去时态。通常情况下，展现数据图表时用现在时态，描述假设和原理时也用现在时态。介词的正确使用，不光对母语非英语的科技人士来说是难题，对母语为英语的人来说也很难弄清楚。介词与形容词、动词、名词之间有着广泛的搭配关系。要掌握介词的用法首先应注意这些搭配关系。科技论文写作中常使用的介词搭配：

attempt(n.) **at**　　　　　　　attempt(v.) **to**
refer **to**　　　　　　　　　　different **from**
compare **to/ with**　　　　　theorize **about**
in contrast **to**　　　　　　　implicit **in**
correlated **with**　　　　　　through the decrease **of/in**
similar **to**　　　　　　　　　analogous **to**
in connection **with**　　　　search **for/of**
a comparison **of** A **with** B …/a comparison **between** A **and** B…

Unit 8

IoT Security

Passage A   An Overview of IoT Security

Passage B   Access Control

Passage C   SDN Security Considerations in the Data Center

Passage A　An Overview of IoT Security

Ⅰ. An Introduction

The Internet of Things (IoT) paradigm has gained popularity in recent years. The interconnected device networks can lead to a large number of intelligent and autonomous applications and services that can bring significant personal, professional, and economic benefits, resulting in the emergence of more data centric businesses. IoT devices have to make their data accessible to interested parties, which can be web services, smart phone, cloud resource, etc. Making these data available through the Internet is one thing, doing this in a controlled way, not exposing data to the whole world, is another thing. Therefore, the more objects get linked via the Internet of Things, the greater becomes the possibility of digital mischief or mayhem.

Concerns have been raised that the Internet of Things is being developed rapidly without appropriate consideration of the profound security challenges involved and the regulatory changes that might be necessary. According to the BI (Business Insider) Intelligence Survey conducted in the last quarter of 2014, 39% of the respondents said that security is the biggest concern in adopting Internet of Things technology. In particular, as the Internet of Things spreads widely, cyber attacks are likely to become an increasingly physical (rather than simply virtual) threat.

Applying these same practices or variants of them in the IoT world requires substantial reengineering to address device constraints. Blacklisting, for example, requires too much disk space to be practical for IoT applications. Embedded devices are designed for low power consumption, with a small silicon form factor, and often have limited connectivity. They typically have only as much processing capacity and memory as needed for their tasks. And they are often "headless"—that is, there isn't a human being operating them who can input authentication credentials or decide whether an application should be trusted; they must make their own judgments and decisions about whether to accept a command or execute a task.

IoT security encompasses several layers of abstraction and a number of dimensions. The abstraction levels range from physical layers of sensors, computation and communication, and devices to the semantic layer in which all collected information is interpreted and processed. We expect that a majority of security attacks will occur at the software level because it is currently most popular and can simultaneously cover a large number of devices and processes. From a research point of view, most novel attacks are on physical signals and, in particular, semantic attacks during data processing and decision making steps. It is important to observe that the lowest security at any level and at any dimension determines the overall security.

Researchers have identified privacy challenges faced by all stakeholders in IoT domain, from the manufacturers and app developers to the consumers themselves, and examined the responsibility of each party in order to ensure user privacy at all times. Problems highlighted by the report include:

- User consent - somehow, the report says, users need to be able to give informed consent to data collection. Users, however, have limited time and technical knowledge.
- Freedom of choice - both privacy protections and underlying standards should promote freedom of choice.
- Anonymity - IoT platforms pay scant attention to user anonymity when transmitting data, the researchers note.

II. IoT Security Requirement

The practical realization of IoT requires the development of a number of new versions of platforms and technologies including device and process identification and tracking, sensing and actuation, communication, computational sensing, semantic knowledge processing, coordinated and distributed control, and behavioral, traffic, and user modeling. The realization of IoT subsystems will be subjected to numerous constraints that include cost, power, energy, and lifetime. However, there is a wide consensus that the most challenging of requirements will be security. It is widely acknowledged that the potential for malicious attacks can and will be greatly spread and actuated from the Internet to the physical word. Hence, security of IoT is of essential importance.

There are several factors which need to be taken care of while devising a security solution for the IoT devices. The Security requirements that are expected to be met by the IoT security schemes are as follows.

1. Information Security Requirements

(1) Integrity: An adversary can change the data and compromise the integrity of an IoT system. Thus, integrity ensures that any received data has not been altered in transit.

(2) Information protection: The security and confidentiality of the on-air and stored information should be strictly preserved. It refers to limiting the information access and disclosure to the authorized IoT node, and preventing access by or disclosure to unauthorized ones. For instance, an IoT network should not reveal the sensor readings to its neighbours (if it is configured not to do so).

(3) Anonymity: Anonymity hides the source of the data. This security service helps with the data confidentiality and privacy.

(4) Non-repudiation: Non repudiation is the assurance that someone cannot deny something. An IoT node cannot deny sending a message it has previously sent.

(5) Freshness: It is required to ensure the freshness of each message. Freshness guarantees that the data is very much recent and no old messages have been replayed.

2. Access Level Security Requirements

(1) Authentication: Authentication enables an IoT device to ensure the identity of the peer with which it communicates (e.g. receiver verifies whether received data originated from the correct source or not). It also requires to ensure that valid users get access to the IoT devices and networks for administrative tasks: remote reprogramming or controlling of the IoT devices and networks.

(2) Authorization: It ensures that only the authorized devices and the users get access to the network services or resources.

(3) Access control: Access control is the act of ensuring that an authenticated IoT node accesses only what it is authorized to, and nothing else.

3. Functional Security Requirements

(1) Exception handling: Exception handling confirms that an IoT network is alive and continues serving even in the anomalous situations: node compromise, node destruction, malfunctioning hardware, software glitches, dislocation environmental hazards, etc. Thus it assures robustness.

(2) Availability: Availability ensures the survivability of IoT services to authorized parties when needed despite denial of-service attacks. It also ensures that it has the capability to provide a minimum level of

services in the presence of power loss, failures.

(3) Resiliency: In case a few inter connected IoT devices are compromised, a security scheme should still protect against the attack.

(4) Self organization: An IoT device may fail or run out of energy. The remaining device or collaborator devices should have the ability to be reorganized to maintain a set level of security.

III. Trust, Security and Privacy (TSP)

The question of individual privacy and security within this for the individual becomes more difficult as the complex chain within which the security has been created is infinite and the weakest link defines the overall level of security. With IPv6 there are enough IP addresses to go around for the predicted tens of billions of data points that will form our new world – the question is whether they can all be secured to a level that can ensure individual privacy rights and secure the systems from malicious attacks.

In traditional TCP/IP networks, security is built to protect the confidentiality, integrity and availability of network data. It makes the system reliable and protects the system from malicious attacks which can lead to malfunctioning systems and information disclosure. As the characteristic of node and application environment, WSN security not only needs traditional security protection, but also the special requirements of trust, security and privacy WSNs. Trust, security and privacy may, depending on the application scenario, require security protection of integrity, availability, confidentiality, non-repudiation, and user privacy. It supports system integrity, reliability by protecting the system from malicious attacks. TSP WSNs may need to protect the nodes against tampering, protect the communication channel, and routing in the network layer. Logging/audit functions may be required to detect attacks. The technology of TSP WSNs consists of message authentication, encryption, access control, identity authentication, etc.

1. Node Security and Sleep Deprivation

A node of a WSN may be tampered with via its logical interfaces or by direct physical attacks; it may be relocated without authorization, or stolen. Node security may contain secure wakeup and secure bootstrapping. A low duty cycle is crucial to ensure a long lifetime of battery-powered sensor nodes. A special class of denial of service attacks, the so called sleep deprivation attacks prevents the sensor node from going to the power-saving sleep mode, hence severely reduces the lifetime of an attacked sensor node. Standard security mechanisms like message authentication codes or frame encryption do not prevent sleep deprivation attacks: the node is powered up and energy is spent for processing the received message. The attack can only be noticed when battery power has already been spent. The wake-up radio listens on the channel when the sensor node is in sleep state. It triggers the sensor wake up when it receives a wake-up signal. To add security to the general wake-up radio design, the wake-up signal is an encoded wake-up code. As the wake-up code is used only once and as it is specific for each node, it can be sent in clear when waking up a node.

2. Crypto Algorithms

Encryption is a special algorithm to change the original information of the data sensor node, which makes an unauthorized user not recognize the original information even if he has accessed the encrypted information. The IoT of public infrastructure are inevitably exposed to the scope of public activities. Traditional message authentication code, symmetric encryption and public-key encryption have exposed their shortcomings. Therefore an encryption system, which is more suitable for WSNs, needs to be proposed.

IEEE 802.15.4 sets the encryption algorithm to use when cyphering the data to transmit. The encryption algorithm used is AES (Advanced Encryption Standard) with a 128bit key length (16 B). It is really important to count with an unique kind of encryption method due to the fact that most of the 802.15.4/ZigBee transceivers have a specific hardware design to cope with this work at the electronic level (embedded low resources devices).

The AES algorithm is not only used to encrypt the information but to validate the data which is sent. This concept is called Data Integrity and it is achieved using a Message Integrity Code (MIC) also named as Message Authentication Code (MAC) which is appended to the message. This code ensures integrity of the MAC header and payload data attached. It is created encrypting parts of the IEEE MAC frame using the Key of the network, so if we receive a message from a non trusted node we will see that the MAC generated for the sent message does not correspond to the one what would be generated using the message with the current secret Key, so we can discard this message. The MAC can have different sizes: 32, 64, 128 bit, however it is always created using the 128bit AES algorithm. Its size is just the bits length which is attached to each frame. The more large the more secure (although less payload the message can take). Data Security is performed encrypting the data payload field with the 128b Key. The Security in the IEEE 802.15.4 MAC frame is shown as Figure 8.1.

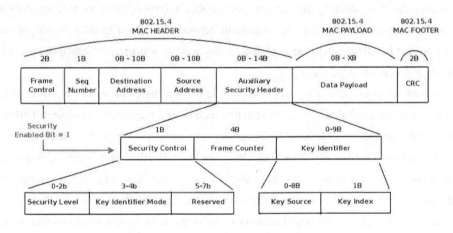

Figure 8.1 Security in the IEEE 802.15.4 MAC frame

3. Key Management

Key management is focused on the area in WSN security. Key management includes key generation, distribution, verification, update, storage, backup, valid and destroy. An effective key management mechanism is also the foundation of other security mechanisms, such as secure routing, secure positioning, data aggregation. Typical key management schemes in WSNs include global key management, random key management, location key management, clustering key management and public key-based key management. The security bootstrapping procedure establishes the security configuration of a sensor node, e.g. a join key is established during the bootstrapping.

As there are multiple bootstrapping procedures and the choice of an appropriate bootstrapping procedure heavily depends on the environment, normal operation of the sensor network is decoupled from the bootstrapping such that it is possible to change the bootstrapping procedure without any change on the security architecture for normal operation. The appropriate bootstrapping procedure depends to a high degree on the application and its environment. Therefore, several different bootstrapping procedures have been proposed: token based, pre-configuration of the keys during manufacturing of the nodes, physical protection of messages, in-band during a weak security set-up phase, out-of-band communication.

4. Secure Data Aggregation

Secure data aggregation is to ensure each node data secure. Therefore, the general processes of secure data aggregation are as follows: first nodes should be possible to provide reliable date and securely transmit them to the higher aggregation nodes. The higher aggregation nodes judge the credibility of data and do aggregation calculation based on redundancy. Each aggregation nodes select the next safe and reliable hop, transmit data to the central node. The central nodes judge the credibility of data and do the final aggregation calculation. Initially, data aggregation regarded energy as the object and barely considered security issues. Now secure data aggregation is mostly realized by authentication and encryption based on the theory of cluster, ring, and hierarchical.

5. Secure Routing

Since WSNs using multi-hop in data transfer and self-organization in networking, each node also needs routing discovery, routing establishment, routing maintenance. Secure routing protocol is that complete effective routing decisions and may be a prerequisite for data aggregation and redundancy elimination safe from a source node to a sink node. Many secure routing networks have been specifically designed for WSNs, they can be divided into three categories according to the network structure: flat-based routing, hierarchical based routing, and location-based routing.

Typical methods of secure routing protocols include methods based on feedback information, location information, encryption algorithm, multipath selection method and hierarchical structures. Different secure routing protocols can solve problems of different types of attacks, such as the secure routing protocol based on the feedback information that includes the information of delay, trust, location, excess capacity in acknowledgment frame of the media access control (MAC) layer. Although not using encryption, this method can resist common attacks such as false routing information, cesspool attack and wormhole. Most current secure routing protocols assume the sensor network is stationary, so more new secure routing protocols need to be developed to satisfy mobility of sensor nodes.

Ⅳ. New Words

actuation [ˌæktjʊˈeɪʃən]　　　　　n. 冲动，驱使；刺激；行动
potential [pəˈtenʃl]　　　　　　　adj. 潜在的；可能的；势的
scheme [skiːm]　　　　　　　　n. 体制；诡计
integrity [ɪnˈtegrɪtɪ]　　　　　　n. 完整性
anonymity [ænəˈnɪmɪtɪ]　　　　n. 匿名；匿名者；无名之辈

Ⅴ. Phrases

conceptual level　　　　　　　概念层
hand-held embedded　　　　　手持嵌入式
interconnected device networks　互联网络设备

Passage B Access Control

Ⅰ. Overview

Access control is a security technique that can be used to regulate who or what can view or use resources in a computing environment. There are two main types of access control: physical and logical. Physical access control limits access to campuses, buildings, rooms and physical IT assets. Logical access limits connections to computer networks, system files and data.

Ⅱ. Access Control Classification

The four main categories of access control are:

1. Mandatory Access Control

Mandatory access control (MAC) is a security strategy that restricts the ability individual resource owners have to grant or deny access to resource objects in a file system. MAC criteria are defined by the system administrator, strictly enforced by the operating system (OS) or security kernel, and are unable to be altered by end users.

2. Discretionary Access Control

Discretionary access control (DAC) is a type of access control defined by the Trusted Computer System Evaluation Criteria as a means of restricting access to objects based on the identity of subjects and/or groups to which they belong. The controls are discretionary in the sense that a subject with a certain access permission is capable of passing that permission (perhaps indirectly) on to any other subject (unless restrained by mandatory access control).

3. Role-based Access Control

Role-based access control (RBAC) is a method of regulating access to computer or network resources based on the roles of individual users within an enterprise. In this context, access is the ability of an individual user to perform a specific task, such as view, create, or modify a file. Roles are defined according to job competency, authority, and responsibility within the enterprise.

4. Rule-based Access Control

Rules Based Access Control is a strategy for managing user access to one or more systems, where business changes trigger the application of Rules, which specify access changes. Implementation of Rules Based Access Control systems is feasible so long as the number of triggering business events and the set of possible actions that follow those events are both small.

Access control systems perform authorization identification, authentication, access approval, and accountability of entities through login credentials including passwords, personal identification numbers (PINs), biometric scans, and physical or electronic keys.

Access control lets only authorized users to access a resource, such as a file, IoT device, sensor or URL. All modern operating systems limit access to the file system based on the user. For instance, the superuser has wider access to files and system resources than regular users. In the IoT context, access control is needed to make sure that only trusted parties can update device software, access sensor data or command the actuators to perform an operation. Access control helps to solve data ownership issues and enables new business models such as Sensors As a Service, where you might for instance sell temperature sensor data to customers. Access control enables companies to share IoT device data selectively with technology vendors to allow both predictive maintenance and protection of the sensitive data.

IoT presents a unique set of access control challenges due to low power requirements of IoT devices, low bandwidth between IoT devices and the Internet, distributed nature of the system, ad-hoc networks, and the potential for extremely large number of IoT devices. This means that standard authorization models, such as Access Control List (ACL), Role Based Access Control (RBAC), Attribute Based Access Control (ABAC) and similar capability-based systems must be analyzed in depth before applying them to the Internet of Things.

III. Access Control Solution for IoT

Intopalo, which was founded in 2012 on a strong mobile product and software development expertise, proposed an access control solution for IoT There are two ways to implement access control for IoT. In a distributed architecture, an access control server grants access tokens to users, who use them to access the IoT devices directly. In a centralized architecture, the user accesses only cloud-based servers that authorize the request and relay data between the user and the IoT devices. All access control models can be implemented using either a distributed or centralized architecture. Most distributed architectures lend principles from the capability-based access control models, which we discuss below (as shown in Figure 8.2).

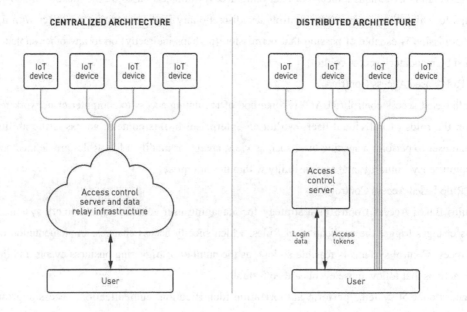

Figure 8.2　Simplistic cloud pictures of access control architectures

The centralized architecture model uses intermediate servers for access control and data relay. This allows for complicated and resource intensive access control models with the low power IoT devices. However, a centralized architecture creates a single point of failure and the privacy of the requester could be compromised. Distributed architectures can scale better and allow a higher degree of privacy for the requester, but it makes complicated access control models such as RBAC and ABAC challenging for low-powered IoT devices.

An access control matrix (as shown in Figure 8.3) is an abstract way to present the access control problem. The resources are in columns and the users are in rows. On a high level, all access control models try to present the access control matrix using an abstraction that makes managing access control on system level easier. Also, most of real-life access control systems are hybrid systems using components from several access control models. For example, the Linux file system is mainly based on access control lists, but it also uses the concept of groups that closely resembles roles in RBAC.

	File 1	File 2	...	File M
User 1			x	
User 2	x	x	...	x
...			x	
User N			x	

The blue x:s demonstrate capabilities associated with a user while red x:s illustrate access control list tied to a resource.

Figure 8.3 access control matrix

An access control list stores the authorized users to each resource. That is, the access control matrix is stored row-wise. For Internet of Things, this presents a big challenge, since devices have low processing power and it is hard to update the user list in a distributed system. Also, the extremely high number of IoT devices and users make fine-grained access control almost impossible using the ACL model.

In the capability-based access control model, the access control matrix is also stored row-wise. All users are given unforgeable access tokens to allow the user to access a resource. In this model, updating a user's access rights doesn't require updating resources, which is a benefit in the IoT context. The majority of the published IoT capability models include an access control server that grants user access tokens which the users can present as they connect directly to the IoT devices.

Role Based Access Control describes the access control matrix in an abstract way. In RBAC, users are associated with roles, which have permissions and therefore the user has a set of permissions based on the roles the user is associated with. Each resource is also associated with permission and access control decision is made based on matching user's permissions against permission associated with the resource. The basic RBAC model can be extended in various ways, for example by associating the roles with dynamic constraints that deactivate the roles if the requirements set forth by the constraints are not met.

The main benefit of RBAC is that adding access rights to users is easy, as long as you can use the existing roles. Downsides include the fact that fine-grained access control leads to role explosion, and that adding new roles usually requires a system wide update. Most (if not all) RBAC implementations have a centralized access control server that grants access to resources. In the IoT context, there is a risk that role explosion becomes hard to handle, especially if fine-grained access control for data ownership is required. Also, system-wide updates might be hard to implement in IoT systems that apply RBAC to the distributed access control architecture.

Attribute Based Access Control describes the access control matrix in an even more abstract way than RBAC. Each user has an access policy that describes user attributes, such as name, job title or responsibilities, and each resource has attributes that describe the conditions that must be met before access is granted. During the authorization, the user attributes are matched against resource's demands and, if all conditions are met, access is granted. Other attributes, such as time or location, can also be considered during the authorization process. In the IoT context, ABAC has many of the same issues as RBAC, since ABAC also works best with a centralized architecture, making system-wide updates more difficult to implement.

The importance of access control in IoT will be emphasized in future, as IoT devices become more widespread and business models become increasingly sophisticated. Perhaps the standard access control models will be eventually replaced by something radically different and better suited for the Internet of Things. For the time being, however, it is beneficial to analyze IoT implementations in the light of the standard access control models in order to tailor an architecture which works best for your product or organization.

Ⅳ. New Words

trigger ['trɪgə]	vt.	引发，引起；触发
superuser [su:pə'ru:zə]	n.	[计]超级用户
tokens ['təʊk(ə)n]	n.	许可证 令牌
intermediate [ˌɪntə'mi:dɪət]	n.	中间物；媒介

Ⅴ. Abbreviations

MAC	Mandatory Access Control	强制访问控制
DAC	Discretionary Access Control	自主访问控制
OS	Operating System	操作系统
RBAC	Role-based Access Control	基于角色的访问控制
PINs	personal Identification Numbers	个人识别号
URL	Uniform Resource Locator	统一资源定位器

Passage C SDN Security Considerations in the Data Center

Secure networks are critical to all businesses, especially with their increased migration to the cloud and the wave of innovation being unleashed by Software-Defined Networking (SDN). SDN provides a centralized intelligence and control model that is well suited to provide much-needed flexibility to network security deployments. OpenFlow™—the first SDN standard—manipulates the network path based on traffic analysis and statistics provided by the SDN controller, in a multi-tenant environment. The flow-based paradigm is particularly well suited to protect traffic on each virtual network slice, or for each virtual tenant. Along with many benefits, SDN poses new threats, particularly with the emergence of cloud, BYOD (Bring Your Own Device), and virtualized environments. It is critical to consider threats, risk exposure, operational impact, performance, scale, and compliance in the SDN-based data center of the future. To illustrate how SDN can be used to improve network security, this section presents a use case for automated malware quarantine (AMQ). AMQ detects and isolates network devices that have become compromised before they can negatively affect the network.

Ⅰ. SDN Overview

Software Defined Networking is a new architecture that has been designed to enable more agile and cost-effective networks. The Open Networking Foundation (ONF) is taking the lead in SDN standardization, and has defined an SDN architecture model as depicted in Figure 8.4.

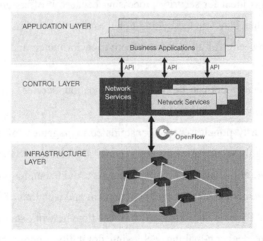

Figure 8.4 SDN architecture

The ONF/SDN architecture consists of three distinct layers that are accessible through open APIs:
- The Application Layer consists of the end-user business applications that consume the SDN communications services. The boundary between the Application Layer and the Control Layer is traversed by the northbound API.
- The Control Layer provides the consolidated control functionality that supervises the network

forwarding behavior through an open interface.
- The Infrastructure Layer consists of the network elements (NE) and devices that provide packet switching and forwarding. According to this model, an SDN architecture is characterized by three key attributes:
- Logically centralized intelligence. In an SDN architecture, network control is distributed from forwarding using a standardized southbound interface: OpenFlow. By centralizing network intelligence, decision-making is facilitated based on a global (or domain) view of the network, as opposed to today's networks, which are built on an autonomous system view where nodes are unaware of the overall state of the network.
- Programmability. SDN networks are inherently controlled by software functionality, which may be provided by vendors or the network operators themselves. Such programmability enables the management paradigm to be replaced by automation, influenced by rapid adoption of the cloud. By providing open APIs for applications to interact with the network, SDN networks can achieve unprecedented innovation and differentiation.
- Abstraction. In an SDN network, the business applications that consume SDN services are abstracted from the underlying network technologies. Network devices are also abstracted from the SDN Control Layer to ensure portability and future-proofing of investments in network services, the network software resident in the Control Layer.

II. The Implications of SDN on Network Security

Open Flow-based SDN offers a number of attributes that are particularly well suited for implementing a highly secure and manageable environment:
- The flow paradigm is ideal for security processing because it offers an endto-end, service-oriented connectivity model that is not bound by traditional routing constraints.
- Logically centralized control allows for effective performance and threat monitoring across the entire network.
- Granular policy management can be based on application, service, organization, and geographical criteria rather than physical configuration.
- Resource-based security policies enable consolidated management of diverse devices with various threat risks, from highly secure firewalls and security appliances to access devices.
- Dynamic and flexible adjustment of security policy is provided under programmatic control.
- Flexible path management achieves rapid containment and isolation of intrusions without impacting other network users. By blending historical and real-time network state and performance data, SDN facilitates intelligent decision-making, achieving flexibility, operational simplicity, and improved security across a common infrastructure.

The SND security policy mangemnt is shown as Figure 8.5.

III. SDN Security Use Case: Automated Malware Quarantine

In this section, we will examine a particular use case to illustrate the implications of the SDN architecture on the implementation of a security function. Automated malware quarantine (AMQ) (as shown in Figure 8.6) detects and isolates insecure network devices before they can negatively affect the network. Upon discovering a potential threat, AMQ identifies the problem and automatically downloads the necessary patches to resolve it. After the threat has been contained, AMQ software automatically allows the device to rejoin the network. This active approach contains and eliminates security threats that could not normally be handled by any single portion of the network. Today, AMQ is typically deployed as a proprietary solution where each device performs its specified function autonomously with limited awareness of other devices in the network. Such approaches are designed for static traffic flows, and must be capable of monitoring in real-time ingress traffic. This closed approach is inflexible, especially for the data center where server workloads are virtualized, traffic flows are highly dynamic, and multiple simultaneous policies must be maintained. Deploying higher-speed links (40G, 100G, etc.) makes this environment even more difficult.

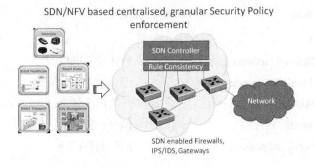

Figure 8.5 SDN Security Policy management

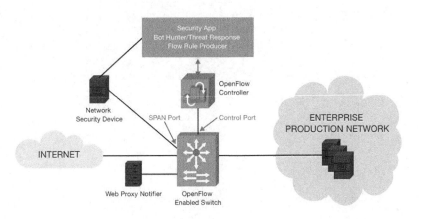

Figure 8.6 Automated malware quarantine

In this example, there are two primary security Network Services Modules (NSMs) hosted on the controller that provide the AMQ function:
- The Bot Hunter NSM monitors the network and detects a malware-infested host in real time.
- The Threat Responder NSM directs the controller to initiate the quarantine procedure to isolate the threat from the network in the event of a malware attack.

AMQ transparently and dynamically applies policies to an individual switched port based on the device or user accessing the port. The automatic reconfiguration reduces the response time to security threats and removes the need to have a network engineer create and apply a policy (VLAN, ACL) to manage network access. This approach minimizes the need for manual configuration and application of network user policies. AMQ does not require any additional network software or hardware beyond the basic OpenFlow-enabled switch or network element, so it is fully interoperable across vendor implementations. AMQ has the potential to reduce operating expenses, automate configuration of edge-port security parameters, and allow for mobility of users at the edge of the network.

IV. New Words

automated ['ɔ:təumeitid]   adj. 自动化的；机械化的
End-User ['end'ju:zə]   n. 终端用户

V. Phrases

physical configuration   实体组态；物理完形

VI. Abbreviations

SDN   Software-Defined Networking   软件定义网络
AMQ   automated malware quarantine   恶意软件自动隔离
ONF   Open Networking Foundation   开放网路基金会
NE   network elements   网络元素
NSM   Network Services Modules   网络服务模块

Exercises

1. 合理的物联网安全架构和安全体系将对物联网的安全使用和可持续发展有着重大影响。

2. 物联网是计算机、互联网与移动通信网等相关技术的演进和延伸。

3. As Internet of things network has expanded rapidly and it's communication network environment has become more complex, the security issues are more complex than the existing network systems.

4. The security measure should have the ability to protect perception layer, network layer, application layer of IOT, cope with the security threats of against IOT, and implement the management of all security measure.

参考译文 Passage A 物联网安全

Ⅰ.概述

近几年,物联网模式已经普及。互联网络设备产生了大量智能和自主的应用和设备,带来了重要的个人的、专业的以及经济方面的效益,也促成了越来越多的以数据为中心的商业模式。物联网设备可以将其数据传输给其他感兴趣的实体,包括网络服务、智能手机、云资源等。将数据传输到互联网是一回事,而用一种控制的方式对这些数据进行传输,且不公开数据是另一回事。因此,越多的对象通过物联网进行连接,就越有可能导致数据损害和混乱。

需要担心的问题是,物联网快速发展,但是对涉及到的安全挑战以及必要的监管改革缺乏适当的考虑。根据2014年最后一季度的财经内幕情报调查,39%的受访者表示,安全问题是采用物联网时最应该关注的问题。尤其是物联网传播十分广泛,网络攻击可能会变成物理的(而不是简单的虚拟性的)威胁。

在物联网中使用传统的安全机制需要重新设计且受到很多限制。例如,对物联网的实际应用来说部署黑名单需要过多的空间。嵌入式设备设计需要考虑低功耗,小型化因素以及有限的连通性。它们通常只有处理它们自身任务所需的能力和内存。同时,它们经常"无领导者",也就是说没有一个人可以输入身份认证或决定是否应该信任应用程序;它们必须做出自己的判断,决定是否接受一个命令或执行一个任务。

物联网安全包含多个抽象层面及维度。抽象层面的范围从物理层面的传感器、计算与通信、设备到语义层面,这一层面是将所收集到的信息进行解释并处理。我们预计,大多数安全袭击都会出现在软件层面,因为其是目前最流行的,并且覆盖的设备和流程也是最广的。从研究角度来看,大多数小型攻击基于物理信号,尤其是数据处理和决策步骤的语义攻击。重要的是要知道,任何层面或任何维度下最低的安全决定了总体安全。

研究人员已经确定,从制造商、软件开发商、到消费者本身,所有物联网领域的利益相关者都面临隐私挑战,需要他们检查每个部分的责任,以确保用户在任何时候的隐私安全。报告反映出来的问题包括:

- 用户同意——报告指出,不管怎样,用户需要能够给予信息收集知情同意。然而用户的时间和技术有限。
- 自由选择——隐私保护和基础标准都应该提升自由选择。
- 匿名——研究人员指出,物联网平台很少关注用户匿名传输数据。

Ⅱ.物联网的安全需求

事实上,物联网的开发需要大量新版本的平台和技术,包括设备和过程辨识、跟踪、传感和驱动、通信、计算感应、语义知识处理、协调、分布式控制、行为、流量以及用户建模。物联网子系统的实现受到很多因素的限制,包括成本、电力、能源以及寿命。然而,一个广泛的共识是,最具有挑战的需求是安全。人们普遍认为,从互联网到物理世界将会广泛地传播潜在的恶意攻击。因此,物联网的安全极其重要。

在为物联网设备制定安全解决方案时,应该考虑到多个因素。人们期望的能满足安全需求的物联网安全方案如下。

1. 信息安全需求

（1）完整性：破坏者可以改变数据，破坏物联网系统的完整性。因此，完整性确保了任何接收到的数据在传输过程中不会发生改变。

（2）信息保护：广播和储存信息的保密性和机密性都应该保护。它是指限制信息访问或披露给授权的物联网节点，并且阻止未授权的节点访问信息。例如，一个物联网网络不应该将其传感器读数提供给相邻网络（如果是配置不允许）。

（3）匿名：匿名隐藏数据的来源。这个安全服务有助于数据保密性和隐私。

（4）不可否定性：非否定性保证有人不能否定一些东西。一个物联网节点不能否认之前发送过的消息。

（5）实时性：应确保每条信息的实时性。实时性保证了数据是最近的，并且不会存在旧信息重播。

2. 访问级的安全需求

（1）认证：认证可以使物联网设备保证通信之间的设备的身份（例如，接收器可以识别接收到的数据来源是否正确）。还需要保证合法用户可以使用物联网设备及网络进行任务管理：远程改变或控制物联网的设备和网络。

（2）授权：确保只有授权的设备和用户，才能获得网络服务或资源。

（3）访问控制：访问控制要确保授权的物联网节点只能访问其被授权的，而没有授权的不能访问。

3. 功能性的安全需求

（1）异常处理：异常处理可以确认物联网网络可用，并能在异常情况下继续提供服务，节点妥协，节点破坏，硬件故障，软件故障，位错环境危害等。因此，其能保证鲁棒性。

（2）可用性：可用性确保物联网在受到DoS攻击时，仍然能够生存。它有能力在电量不足和失效时提供最低的服务水平。

（3）可恢复性：如果几个物联网设备被破坏,安全方案应该阻止攻击。

（4）自组织：物联网设备可能会失效或耗尽能量。剩下的设备或合作设备要有重组的能力去维护安全。

III. 可信、安全和隐私

对于个人而言，个人隐私和安全问题变得越来越困难，这是由于已有的安全性没有限制，且最弱的一环被用于定义整体安全水平。IPv6有足够的IP地址用于预测数百亿的数据点，这将会形成一个全新的世界——问题在于他们是否能全部在一个水平，保证个人的隐私权，确保系统不会遭到恶意攻击。

在传统的TCP/IP网络中，建立安全机制是用于保护网络数据的机密性、完整性和可用性。它使系统具有可靠性并且避免系统遭到恶意攻击，这种攻击会导致系统故障和信息披露。作为节点和应用环境的特点，无线传感器网络安全不仅需要传统的安全保护，而且还要求特别的信任、安全及隐私保护需求。可信、安全和隐私可能会取决于应用场景，它需要保护信息的完整性、可用性、保密性、不可否认性和用户隐私。它通过保护系统免受恶意攻击，支持系统的完整性，可靠性。可信、安全和隐私的无线传感器网络可能需要保护节点不受干扰，保护通信信道，以及网络层的路由途径。记录和审计功能可能需要检测攻击。可信、安全和隐私的无线传感器网络技术包括消息认证、加密、访问控制、身份认证等。

1. 节点安全和休眠损耗

传感器网络的节点可能会通过逻辑接口或直接的物理性攻击，受到干扰；在没有授权或者被盗的情况下，其可能会被重新安装。节点安全包括安全唤醒和引导。低占空比非常重要，用于确保电池驱动的传感器节点有较长的寿命。拒绝服务攻击的特殊类型被称为休眠损耗攻击，用于阻止传感器节点进入节电休眠模式，因此会严重降低已受攻击传感器节点的寿命。标准的安全机制就像信息验证码，或者帧加密，不会阻止休眠损耗攻击：启动节点，用能量处理接收到的信息。攻击只有当电池耗完时才会被注意到。唤醒的无线电广播设备，可以在传感器节点处于休眠状态时，监听通道。当接收到唤醒信号时，它触发传感器唤醒。增加唤醒无线广播的安全，唤醒信号必须是一个编码的唤醒代码。唤醒代码只能被使用一次，并且特定于每个节点，在唤醒节点时被明确发送。

2. 加密算法

加密是一种特殊的算法，用于改变数据传感器节点的原始信息，这使得未经授权的用户即使访问加密信息也不能识别原始信息。物联网的公共基础设施，不可避免地会暴露公共活动的范围。传统的信息认证码、对称加密及公开密钥加密都会暴露他们的不足之处。因此，需要提出更适合无线传感器网络的加密系统。

IEEE 802.15.4规定的加密算法用于传输中的数据加密。采用的加密算法是AES（高级加密标准），拥有128位的密钥长度（16字节）。特殊的加密算法非常重要，由于大部分的802.15.4/无线个域网收发器，都有一个特定的硬件设计，用来处理电子级的工作（嵌入式低资源设备）。

AES算法不仅用于加密信息，也能验证发送的数据。这个概念被称为数据完整性，通过信息完整性校验码实现（MIC），也称为信息验证码（MAC），这被附加到消息上。此代码会确保信息验证码数据头和附属的有效负载数据的完整性。这是使用网络密钥创建IEEE MAC帧的加密部分，因此如果接收到的信息来自于非信任的节点，可以看到为发送信息产生的信息验证码，与使用当前密钥信息产生的不一致，因此我们就可以放弃这条信息了。信息验证码有不同的大小：32、64、128位，一般由128位AES算法生成。验证码的大小就是比特长度，也与每一帧相联系。验证码更大就更安全（尽管携带较少的有效负载）。数据安全通过加密128位密钥的数据负载字段来实现。IEEE 802.15.4 MAC帧的安全如图8.1所示。

3. 密钥管理

密钥管理集中在无线传感器网络安全领域。密钥管理包括密钥的生成、分布、验证、更新、存储、备份、认证及毁坏。有效的密钥管理机制是其他安全机制的基础，例如，安全路由、安全定位及数据聚合等。无线传感器网络中典型的密钥管理模式包括全局密钥管理、随机密钥管理、本地化密钥管理、分类密钥管理及基于公钥的密钥管理。安全引导程序建立了传感器节点的安全结构，例如，连接密钥就是在引导时建立的。

因为多个引导程序及适当引导程序的选择，在很大程度上都取决于环境因素，所以传感器网络的正常操作从引导中解耦了，以至于在普通操作安全结构没有改变的情况下，改变引导程序也是可能的。适当的引导程序很大程度上取决于应用程序以及环境因素。因此，几种不同的引导程序被提出：标识符、节点制造过程中对关键节点的预配置、信息的实体保护、较弱安全设置阶段带内以及带外通信。

4. 安全数据聚合

安全数据聚合是为了确保各个节点数据的安全。因此，安全数据聚合的一般过程是：首先，节点提供可靠的数据并安全地把其传输到高聚合节点。高聚合节点可判断数据的可信度，并能基于冗余进行聚合计算。各个聚合节点都会选择下一个安全并可靠的跳行,传输数据给中心节点。中心节点可判断数据的可信度,并完成最后的聚合计算。最初,数据聚合将能量视为对象,几乎不考虑安全问题。而现在,安全数据聚合大多数是通过认证和加密来实现,认证和加密基于聚合、环形和分层理论。

5. 安全路由

由于无线传感器网络在数据传送时使用多跳,在网络中有自我组织,各个节点需要路由发现、路由建立以及路由维护。安全路由协议是完全有效的路由决策,并可能是从源节点到汇聚点数据聚合与安全冗余消除的先决条件。许多安全路由网络都是为无线传感器网络专门设计的,根据网络结构,可以被分为三类:平面路由、基于分层的路由以及基于位置的路由。

安全路由协议的典型方法包括基于反馈信息、位置信息、加密算法、多路径选择方法和层次结构法。不同的安全路由协议可以解决不同类型的攻击,例如,基于反馈信息的安全路由协议,包括信息延迟、信任、位置以及关于媒体访问控制(信息验证码)层确认帧的相关能力。虽然没有使用加密技术,该方法可以抵抗常见的攻击,例如,虚假路由信息、污水沟攻击和虫洞攻击。目前,多数路由协议都假设传感器网络是固定的,因此需要开发更多新的安全路由协议,以满足传感器节点的移动性。

参考答案

1. A well-defined security architecture and security system for the IOT will have great impact on the save application and sustainable development of the IOT.

2. Internet of Things (IoT) is seen as the evolution of related technologies such as computer, Internet and mobile networks.

3. 随着物联网网络规模迅速扩大，物联网的网络环境变得越来越复杂，物联网的安全问题将比现有网络系统更加复杂和难以解决。

4. 物联网安全需要对物联网感知层、网络层和应用层进行有效的安全保障，以应对其面临的安全威胁，并且还要能够对各个层次的安全防护进行管理。

物联网专业信息检索

在我们的专业学习过程中，需要查阅大量的外文文献，来补充自己的不足或增加对知识点的了解，这里列举几个文献检索数据库供读者参考。

1. IEEE/IEL 数据库

IEL 数据库(IEEE/IET Electronic Library)是 IEEE 旗下最完整的在线数据资源，它提供了当今世界在电气工程、通信工程和计算机科学领域中近三分之一的文献，并在多个学科领域引用量名列前茅。IEL 数据库内容包含：170种 IEEE 的期刊、会刊与杂志（最早回溯到1893年）；20多种 IET 的期刊；每年超过1200种 IEEE 的会议录；每年20多种 IET/VDE 会议录；超过2400种 IEEE 现行、存档标准；所有文献的 Inspec 索引目录。

2. EBSCO 数据库（综合学科，期刊，远程）

EBSCO 是一个具有60多年历史的大型文献服务专业公司，提供期刊、文献定购及出版等服务，总部在美国，在19个国家设有分部。开发了近100多个在线文献数据库，涉及自然科学、社会科学、人文和艺术等多种学术领域。其中两个主要全文数据库是：Academic Search Premier 和 Business Source Premier。学术期刊集成全文数据库（Academic Search Premier）总收录期刊7699种，其中提供全文的期刊有3971种，总收录的期刊中经过同行鉴定的期刊有6553种，同行鉴定的期刊中提供全文的有3123种，被 ISCI & SSCI 收录的核心期刊为993种（全文有350种）。主要涉及工商、经济、信息技术、人文科学、社会科学、通信传播、教育、艺术、文学、医药、通用科学等多个领域。

3. Web of Science 会议录引文索引

Web of Science 是 ISI 数据库中的引文索引数据库，共包括8000多种世界范围内最有影响力的、经过同行专家评审的高质量的期刊。是大型综合性、多学科、核心期刊引文索引数据库，包括三大引文数据库[科学引文索引（Science Citation Index，SCI）、社会科学引文索引（Social Sciences Citation Index，SSCI）和艺术与人文科学引文索引（Arts & Humanities Citation Index，A&HCI）]和两个化学信息事实型数据库（Current Chemical Reactions，CCR 和 Index Chemicus，IC），以及科学引文检索扩展版（Science Ciation Index Expanded, SCIE）、科技会议文献引文索引（Conference Proceedings Citation Idex-Science, CPCI-S）和社会科学以及人文科学会议文献引文索引（Conference Proceedings Citation index-Social Science&Humanalities, CPCI-SSH）3个引文数据库，以 ISI Web of Knowledge 作为检索平台。

其他数据库还包括：

- Springer 电子图书；
- SIAM（工业和应用数学学会）全文期刊数据库；
- ElsevierSD（工程类、计算机类期刊）；
- ACM 全文电子期刊数据库（计算机）；
- Emerald 全文期刊库；
- Derwent Innovations Index 德温特专利创新索引；
- Emerald 回溯内容数据库；
- Springerlink 数据库；
- Ei Village 工程索引。

Unit 9

Cloud Computing

Passage A   An Overview of Cloud Computing

Passage B   Mobile Cloud Computing and Granules Runtime for Cloud Computing

Passage A   An Overview of Cloud Computing

Ⅰ. An Introduction

Cloud computing has been envisioned as the next generation computing model for its major advantages in on demand self-service, ubiquitous network access, location independent resource pooling and transference of risk. Cloud computing is the latest developments of computing models after distributed computing, parallel processing and grid computing. Cloud computing achieve multi-level virtualization and abstraction through effective integration of variety of computing, storage, data, applications and other resources, users can be easy to use powerful computing and storage capacity of cloud computing only need to connect to the network. The concept of cloud computing is shown as Figure 9.1.

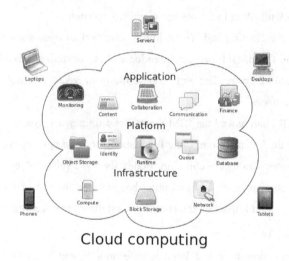

Figure 9.1   Cloud Computing

The present availability of high-capacity networks, low-cost computers and storage devices as well as the widespread adoption of hardware virtualization, service-oriented architecture, and autonomic and utility computing have led to a growth in cloud computing. Companies can scale up as computing needs increase and then scale down again as demands decrease.

Cloud computing has become a highly demanded service or utility due to the advantages of high computing power, cheap cost of services, high performance, scalability, accessibility as well as availability. Some cloud vendors are experiencing growth rates of 50% per year, but being still in a stage of infancy, it has pitfalls that need to be addressed to make cloud computing services more reliable and user friendly.

Ⅱ. The History of Cloud Computing

The origin of the term cloud computing is unclear. The word "cloud" is commonly used in science to describe a large agglomeration of objects that visually appear from a distance as a cloud and describes any set of things whose details are not further inspected in a given context. Another explanation is that the old programs that drew network schematics surrounded the icons for servers with a circle, and a cluster of servers in a network diagram had several overlapping circles, which resembled a cloud. References to "cloud computing" in its modern sense appeared as early as 1996, with the earliest known mention in a Compaq internal document.

The popularization of the term can be traced to 2006 when Amazon.com introduced its Elastic Compute Cloud.

In the 1990s, telecommunications companies, who previously offered primarily dedicated point-to-point data circuits, began offering virtual private network (VPN) services with comparable quality of service, but at a lower cost.

In early 2008, NASA's OpenNebula, enhanced in the RESERVOIR European Commission-funded project, became the first open-source software for deploying private and hybrid clouds, and for the federation of clouds. In the same year, efforts were focused on providing quality of service guarantees (as required by real-time interactive applications) to cloud-based infrastructures, in the framework of the IRMOS European Commission-funded project, resulting in a real-time cloud environment.

In July 2010, Rackspace Hosting and NASA jointly launched an open-source cloud-software initiative known as OpenStack. The OpenStack project intended to help organizations offering cloud-computing services running on standard hardware. The early code came from NASA's Nebula platform as well as from Rackspace's Cloud Files platform.

On March 1, 2011, IBM announced the IBM SmartCloud framework to support Smarter Planet. Among the various components of the Smarter Computing foundation, cloud computing is a critical part.

According to the top ten strategic technology trends for 2012 provided by Gartner (a famous global analytical and consulting company), cloud computing has been on the top of the list, which means cloud computing will have an increased impact on the enterprise and most organizations in 2012.

III. Cloud Computing Architecture

We consider the cloud computing is a large scale economic and business computing paradigm with virtualization as its core technology. The cloud computing system is the development of parallel processing, distributed and grid computing on the Internet, which provides various QoS guaranteed services such as hardware, infrastructure, platform, software and storage to different Internet applications and users.

1. Framework

Cloud computing systems actually can be considered as a collection of different services, Cloud computing is broken down into three segments: applications, platforms and infrastructure (as shown in Figure 9.2). Each segment serves a different purpose and offers different products for businesses and individuals around the world.

(1) Infrastructure layer: it includes resources of computing and storage.

(2) Platform layer: this layer is considered as a core layer in the cloud computing system, which includes the environment of parallel programming design, distributed storage and management system for structured mass data, distributed file system for mass data, and other system management tools for cloud computing.

(3) Application layer: this layer provides some simple software and applications, as well as costumer interfaces to end users.

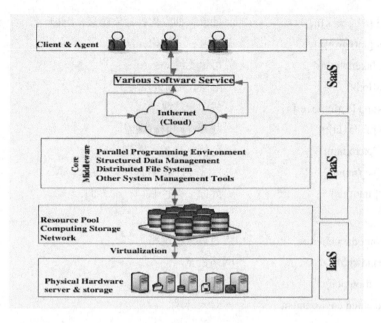

Figure 9.2 The Framework of Cloud Computing

2. Features

The features of Cloud Computing are as follows:

(1) Virtualization: the "Cloud" can be considered as a virtual resource pool where all bottom layer hardware devices is virtualized.

(2) Reliability: usability and extensibility: cloud computing provides a safe mode to store user's data while users do not worry about the issues such as software updating, leak patching, virus attacks and data loss.

(3) Large-scale: in order to possess the capability of supercomputing and mass storage, a cloud computing system normally consists of thousands of servers and PCs.

(4) Autonomy: a cloud system is an autonomic system, which automatically configures and allocates the resources of hardware, software and storage to clients on-demand, and the management is transparent to end users.

3. Service Model

(1) Software-as-a-Service (SaaS): Cloud application services or Software as a Service is software that is deployed over the internet and/or is deployed to run behind a firewall in our local area network or personal computer.

(2) Platform-as-a-Service (PaaS): Cloud platform services or Platform as a Service, another SaaS and this kind of cloud computing provide development environment as a service.

(3) Infrastructure-as-a-Service (IaaS): Cloud infrastructure services or Infrastructure as a service delivers a platform virtualization environment as a service.

Ⅳ. New Words

ubiquitous [juːˈbɪkwɪtəs]   adj. 泛在，无所不在的；普遍存在的
virtualization [vɜːtʃʊəlaɪˈzeɪʃn]   n. 虚拟化

abstraction [æb'strækʃn]　　　　　　n. 抽象；抽象化；抽象概念；出神
integration [,ɪntɪ'greɪʃn]　　　　　　n. 整合；一体化；结合
scalability [skeɪlə'bɪlɪtɪ]　　　　　　n. 可量测性
pitfalls ['pɪtfɔ:lz]　　　　　　　　n. 陷阱，易犯的错误
agglomeration [ə'glɒmə'reɪʃn]　　　n. 成团，结块
federation [,fedə'reɪʃn]　　　　　　n. 联邦，同盟；联盟
paradigm ['pærədaɪm]　　　　　　　n. 范例，样式，模范
segments [seg'mənts]　　　　　　　n. [计算机]（字符等的）分段；瓣
interfaces ['ɪntəfeɪs]　　　　　　　n. 界面；<计>接口；交界面

Ⅴ. Phrases

point-to-point data circuits　　　　点对点数据电路
network bandwidth　　　　　　　网络带宽
the demarcation point　　　　　　分界点
a real-time cloud environment　　　实时云环境
cloud computing architecture　　　云计算架构
service model　　　　　　　　　服务模型

Ⅵ. Abbreviations

VPN　　virtual private network　　虚拟专用网

Passage B Mobile Cloud Computing and Granules Runtime for Cloud Computing

Ⅰ. **Mobile Cloud Computing**

Mobile Cloud Computing (MCC) paradigm is the intersection between both mobile and cloud computing, as illustrated by Figure 9.3. it overlaps both technologies yet it presents its own uniqueness over both paradigms. In particular, MCC allows developers to create applications with enhanced cloud services that have fewer limitations over native mobile applications. However, MCC comes with higher complexity that affects the not only the reliability and delivery of its application but also its development lifecycle.

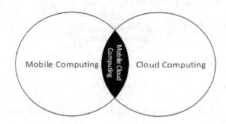

Figure 9.3 Mobile Cloud Computing

As an inheritance and development of cloud computing, resources in mobile cloud computing networks are virtualized and assigned in a group of numerous distributed computers rather than in traditional local computers or servers, and are provided to mobile devices such as smartphones, portable terminal, and so on.

Mobility has become a very popular word and rapidly increasing part in today's computing area. An incredible growth has appeared in the development of mobile devices such as, smartphone, PDA, GPS Navigation and laptops with a variety of mobile computing, networking and security technologies. In addition, with the development of wireless technology like WiMax, Ad Hoc Network and Wi-Fi, users may be surfing the Internet much easier but not limited by the cables as before. Thus, those mobile devices have been accepted by more and more people as their first choice of working and entertainment in their daily lives.

Ⅱ. **Granules Runtime for Cloud Computing**

The Granules project is a lightweight streaming-based runtime for cloud computing. Granules orchestrates the concurrent execution of applications on multiple machines. The runtime manages an application's execution through various stages of its lifecycle: deployment, initialization, execution and termination.

The sheer scale of the computing and storage capabilities available in datacenters make cloud computing particularly well-suited for scientific applications. These scientific applications can be compute-intensive, data-intensive or both: in each case the processing needs are significant enough to mandate concurrent processing. It can be argued that cloud computing is a natural evolution of the grid computing paradigm wherein the services are hosted on a compute cloud that can comprise a much larger set of machines.

Granules uses NaradaBrokering as the content distribution network for disseminating streams. At each computational resource, Granules manages the concurrent execution of multiple application instances. Figure 9.4 depicts the components that comprise Granules.

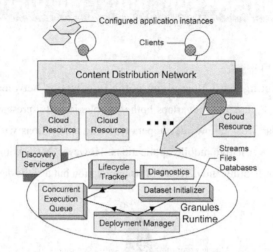

Figure 9.4　Overview of the Granules runtime

One of the characteristics of cloud computing is that there sources are dynamic. If an application programmer is forced to keep track of all the available computational resources, the concomitant increase in the application's complexity would be significant. Granules obviates the need for applications to track resource availability. Instead, applications delegate the responsibility of discovering and harnessing computational resources to Granules. Depending on the requirements, Granules can deploy application instances on one or more computational resources.

Granules performs two key steps during the deployment of the application instance. First, it initializes the state of the application instance based on the specified initialization directives. Second, it initializes the datasets that the application operates on. Granules allows application instances to specify an execution profile that will govern their lifetime and scheduling strategy.

After every scheduled execution, Granules checks to see if the application instance is ready for termination based on the specified execution profile. If it is determined that an application instance is ready for termination, Granules removes it from the concurrent execution queue and reclaims any allocated resources. Reclaiming involves unsubscribing from any of the streams that the application instance was consuming and clearing any file locks that were established. Components that register for diagnostic messages related to task executions are notified about any lifecycle transitions.

Ⅲ. New Words

paradigm ['pærədaɪm]　　　　　n. 范例，样式，模范
intersection [ˌɪntə'sekʃn]　　　　n. 横断，横切；交叉
inheritance [ɪn'herɪtəns]　　　　n. 继承；遗传；遗产
granules [ɡrænjuːls]　　　　　　n. 小颗粒，小硬粒
orchestrates ['ɔːkɪstreɪt]　　　　vt. 精心策划；把（乐曲）编成管弦乐

disseminating [dɪˈsemɪneɪtɪŋ]	v. 散布，传播
initialization [ɪˌnɪʃəlaɪˈzeɪʃn]	n. 设定初值，初始化
termination [ˈtɜːmɪˈneɪʃn]	n. 结束；终止妊娠
executions [ˌeksɪˈkjuːʃənz]	n. 实行

Ⅳ. **Phrases**

development lifecycle	发展生命周期
portable terminal	便携式终端

Ⅴ. **Abbreviations**

MCC	Mobile Cloud Computing	移动云计算
PDA	Personal Digital Assistant	掌上电脑
GPS	Global Positioning System	全球定位系统

Exercises

Translate the following sentences into Chinese or English.

1. Existing scheduling approaches are inadequate for real-time tasks running in uncertain cloud environments, because those approaches assume that cloud computing environments are deterministic and pre-computed schedule decisions will be statically followed during schedule execution.

2. Green cloud computing has become a major concern in both industry and academia, and efficient scheduling approaches show promising ways to reduce the energy consumption of cloud computing platforms while guaranteeing QoS requirements of tasks.

3. 该方案要求密码，特别是公钥基础设施与SSO和LDAP的操作一致，以确保所涉及的数据和通信的合法性、完整性和机密性。

4. 要使云计算实现其潜能，需要清楚地了解所涉及的各种各样的问题，无论从供应商和消费者的角度，还是从技术的角度都要有清楚的认识。

参考译文　Passage A　云计算技术

Ⅰ. 概述

云计算在按需自助服务、随处可用的网络访问、位置独立的资源池和风险转移等方面具有主要优势，被认为是下一代的计算模型。云计算是继分布式计算、并行处理和网格计算之后最新的计算模型。通过有效地集成各种计算、存储、数据、应用和其他资源，云计算实现了多层次的虚拟和抽象，用户只需连接网络，就能方便快捷地使用云计算强大的计算和存储能力。云计算的概念如图9.1所示。

目前可用的高容量网络、低成本计算机和存储设备、硬件虚拟技术的广泛应用、服务导向型的体系结构以及自主通用的计算，都促进了云计算的发展。公司可以随着计算需求增加而扩大，也可以随着计算需求降低而减小。

因为具有高计算能力、廉价服务、高性能、扩展性强、高可达性和高可用性等优势，云计算已经成为一种高需求的服务或实用设施。尽管一些云供应商每年的增长率达到50%，但云计算仍处于初级阶段。另外，为了使云计算服务更加可靠并对用户更友好，还有一些陷阱需要解决。

Ⅱ. 云计算简史

云计算这个术语的起源是不确定的。"云"这个词在科学上通常用来描述一个大物体的聚集，就如在视觉上出现的物体。它也被用来描述任何给定环境中细节不明的事物。另外一个解释是，人们之间画网络示意图的程序在服务器图标的周围画一个圈，而且网络图中的一组服务器就会有好多重叠的圈，从而看起来像云朵。具有现代意义的术语"云计算"最早出现在1996年康派克公司的内部文件中。

此术语的普及可以追溯到2006年，亚马孙公司引进的弹性计算云技术。

在20世纪90年代，先前主要提供专用的点对点数据电路的通讯企业，开始提供虚拟专用网络（VPN）服务，这些服务以更低的成本实现同样的服务质量。

2008年初，NASA的OpenNebula在欧洲委员会资助的项目RESERVOIR中得以提升，成为第一个开放资源的软件来部署私有云、混合云以及云联盟。同年，工作都集中在为基于云的基础设施（实时交互应用所要求的）提供服务质量保证。在欧洲委员会资助的项目IRMOS的架构中，体现为实时云环境。

2010年7月，Rackspace Hosting和NASA联合发布了开放资源的云软件——OpenStack。OpenStack项目打算为组织机构提供在标准硬件上运行的云计算服务。最早的代码，源于NASA的星云平台和Rackspace的云文件平台。

2011年3月1号，为了支持智慧星球，IBM发布了IBM智能云架构。在各种智能计算基础部件中，云计算是很关键的部分。

根据高德纳公司（著名的全球分析咨询公司）提供的2012年十大战略技术趋势，云计算已位于榜首，这意味着云计算在2012年对企业和多数组织机构的影响将持续增长。

Ⅲ. 云计算架构

我们认为，云计算是一个大规模的经济和商业计算范例，以虚拟化作为核心技术。云计算系统随着并行处理、分布式和因特网网格计算的发展而发展，这些技术提供了各种QoS保证服务，如硬件、基础设施、平台、软件以及不同的因特网应用和用户的存储。

1. 架构

实际上，我们可以认为云计算系统是不同服务的聚集，可被分为三层：应用层、平台层和基础设施层（如图9.2所示）。每层具有不同的目标，而且在世界范围内为商业和个人提供不同的产品。

（1）基础设施层：包括计算和存储资源。

（2）平台层：该层是云计算系统的核心层，包括并行处理设计的环境，结构化海量数据的分布式存储和管理系统，海量数据的分布式文件系统以及其他云计算的系统管理工具。

（3）应用层：该层有一些简单的软件、应用以及终端用户的交互界面。

2. 特点

云计算的特点如下。

（1）虚拟化："云"是虚拟资源池，所有底层硬件设备都是虚拟化的。

（2）可靠性、可用性和可扩展性：云计算提供一个安全的模式存储用户数据，而且用户不需要担心软件更新、漏洞修补、病毒攻击和数据丢失等问题。

（3）大规模：为了具备超级计算和海量存储的能力，云计算系统一般由成千上万个服务器和PC组成。

（4）自治性：云系统是一个自治系统，它能够为用户按需自动配置和分配硬件、软件、存储等资源，而且对终端用户透明管理。

3. 服务模型

（1）软件即服务（SaaS）：云应用服务或软件服务是部署在因特网的软件和/或部署在本地局域网或个人计算机的防火墙后运行的软件。

（2）平台即服务（PaaS）：云平台服务或者平台服务，是另外一种SaaS云计算服务，提供开发环境作为服务。

（3）基础设施即服务（IaaS）：云基础设施服务或基础设施服务虚拟化环境提供平台。

参考答案

1. 因为现有的调度方法假设云计算环境是确定的，而且预先计算的调度决策在调度执行过程中将会是静态的，所以，这些方法对于在不确定的云环境中运行的实时任务来说是不合适的。

2. 绿色云计算已经成为业界和学术界关注的重点。而且，有效的调度方法显示，保证服务质量要求的同时，减少云计算平台能源消耗的方法是大有希望的。

3. The proposed solution calls upon cryptography, specifically Public Key Infrastructure operating in concert with SSO and LDAP, to ensure the authentication, integrity and confidentiality of involved data and communications.

4. If cloud computing (CC) is to achieve its potential, there needs to be a clear understanding of the various issues involved, both from the perspectives of the providers and the consumers of the technology.

物联网相关国际学术期刊

目前物联网相关的国际学术期刊有很多，这里列举一些IEEE、Springer及Elsevier出版的物联网方向的国际期刊，供读者参考。

1. IEEE Internet of Things Journal

IEEE Internet of Things (IoT) Journal publishes articles on the latest advances, as well as review articles, on the various aspects of IoT. Topics include IoT system architecture, IoT enabling technologies, IoT communication and networking protocols such as network coding, and IoT services and applications. Examples are IoT demands, impacts, and implications on sensors technologies, big data management, and future internet design for various IoT use cases, such as smart cities, smart environments, smart homes, etc. The fields of interest include: IoT architecture such as things-centric, data-centric, service-oriented IoT architecture; IoT enabling technologies and systematic integration such as sensor technologies, big sensor data management, and future Internet design for IoT; IoT services, applications, and test-beds such as IoT service middleware, IoT application programming interface (API), IoT application design, and IoT trials/experiments; IoT standardization activities and technology development in different standard development organizations (SDO) such as IEEE, IETF, ITU, 3GPP, ETSI, etc.

http://ieeexplore.ieee.org/xpl/aboutJournal.jsp?punumber=6488907

2. IEEE Sensors Journal

The IEEE Sensors Journal is a peer-reviewed, semi-monthly online journal devoted to sensors and sensing phenomena. According to the IEEE Sensors Council's constitution, "The fields of interest of the Council and its activities shall be the theory, design, fabrication, manufacturing and application of devices for sensing and transducing physical, chemical, and biological phenomena, with emphasis on the electronics and physics aspects of sensors and integrated sensor-actuators." The IEEE Sensors Journal focuses on the numerous sensor technologies spanned by the IEEE, and on emerging sensor technologies.

http://ieeexplore.ieee.org/xpl/RecentIssue.jsp?punumber=7361

3. IEEE Transactions on Industrial Informatics

The scope of the journal considers the industry's transition towards more knowledge-based production and systems organization and considers production from a more holistic perspective, encompassing not only hardware and software, but also people and the way in which they learn and share knowledge. The journal focuses on the following main topics: Flexible, collaborative factory automation, Distributed industrial control and computing paradigms, Internet-based monitoring and control systems, Real-time control software for industrial processes, Java and Jini in industrial environments, Control of wireless sensors and actuators, Systems interoperability and human machine interface.

http://tii.ieee-ies.org/

4. International Journal of Wireless Information Networks

The International Journal of Wireless Information Networks examines applications such as sensor and mobile ad-hoc networks, wireless personal area networks, wireless LANs, mobile data networks, location aware networks and services, wireless health, body area networking, cyber physical systems, opportunistic localization for wireless devices and indoor geolocation, and RF localization and RFID techniques. The journal also covers performance-predictions methodologies, radio propagation studies, modulation and coding, multiple access methods, security and privacy considerations, antenna and RF subsystems, VLSI and ASIC design, experimental trials, traffic and frequency management, and network signaling and architecture.

http://link.springer.com/journal/10776

5. Technology and Economics of Smart Grids and Sustainable Energy

The aim of this journal is to identify challenges and solutions for the energy grids and markets of the future. Smart grids are perceived as a key building block for a sustainable energy future with less emissions and better quality of service. It is becoming obvious that smart energy is an issue of technology developments but at the same time more than that since all the technologies have to be embedded in a network. Underlying the physical network there is a network of stakeholders including multiple consumers, producers and grid operators. Regulation also plays an important role as it shapes the relations between the aforementioned actors. Successful smart grid applications will adapt to the local grid and its characteristics, e.g., density of population and consumption, but they will also adapt to and adjust the local institutional settings.

http://link.springer.com/journal/40866

6. Computer Networks

Computer Networks is an international, archival journal providing a publication vehicle for complete coverage of all topics of interest to those involved in the computer communications networking area. The topics covered by the journal include Communication Network Architectures, Communication Network Protocols, Network Services and Applications, Network Security and Privacy and so on.

http://www.journals.elsevier.com/computer-networks/

7. Journal of Network and Computer Applications

The Journal of Network and Computer Applications welcomes research contributions, surveys and notes in all areas relating to computer networks and applications thereof. The following list of sample-topics is by no means to be understood as restricting contributions to the topics mentioned:

http://www.journals.elsevier.com/journal-of-network-and-computer-applications/

8. Digital Communications and Networks

Digital Communications and Networks, fully open accessed by ScienceDirect, publishes rigorously peer-reviewed and high quality original articles and authoritative reviews that focus on communication systems and networks. Special areas of interest include

http://www.journals.elsevier.com/digital-communications-and-networks/

9. Wireless Networks

The wireless communication revolution is bringing fundamental changes to data networking, telecommunication, and is making integrated networks a reality. By freeing the user from the cord, personal communications networks, wireless LAN's, mobile radio networks and cellular systems, harbor the promise of fully distributed mobile computing and communications, any time, anywhere.

Focusing on the networking and user aspects of the field, Wireless Networks provides a global forum for archival value contributions documenting these fast growing areas of interest. The journal publishes refereed articles dealing with research, experience and management issues of wireless networks.

http://link.springer.com/journal/11276

10. Wireless Personal Communications

Wireless Personal Communications is an archival, peer reviewed, scientific and technical journal addressing mobile communications and computing. It investigates theoretical, engineering, and experimental aspects of radio communications, voice, data, images, and multimedia.

A partial list of topics includes propagation, system models, speech and image coding, multiple access techniques, protocols performance evaluation, radio local area networks, and networking and architectures.

The journal features five principal types of papers: full technical papers, short papers, technical aspects of policy and standardization, letters offering new research thoughts and experimental ideas, and invited papers on important and emerging topics authored by renowned experts.

http://link.springer.com/journal/11277

11. JCN

The JOURNAL OF COMMUNICATIONS AND NETWORKS is committed to publishing high-quality papers that advance the state-of-the-art and practical applications of communications and information networks. Theoretical research contributions presenting new techniques, concepts, or analyses, applied contributions reporting on experiences and experiments, and tutorial expositions of permanent reference value are welcome. The subjects covered by this journal include all topics in communication theory and techniques, communication systems, and information networks.

http://www.jcn.or.kr/home/journal/

12. Pervasive and Mobile Computing

Pervasive computing, often synonymously called ubiquitous computing, is an emerging field of research that brings in revolutionary paradigms for computing models in the 21st century. Tremendous developments in such technologies as wireless communications and networking, mobile computing and handheld devices, embedded systems, wearable computers, sensors, RFID tags, smart spaces, middleware, software agents, and the like, have led to the evolution of pervasive computing platforms as natural successor of mobile computing systems. The goal of pervasive computing is to create ambient intelligence where network devices embedded in the environment provide unobtrusive connectivity and services all the time, thus improving human experience and quality of life without explicit awareness of the underlying communications and computing technologies. In this environment, the world around us (e.g., key chains, coffee mugs, computers, appliances, cars, homes, offices, cities, and the human body) is interconnected.

http://www.journals.elsevier.com/pervasive-and-mobile-computing/

13. EURASIP Journal on Wireless Communications and Networking

EURASIP Journal on Wireless Communications and Networking is a peer-reviewed open access journal published under the brand SpringerOpen. The overall aim of the EURASIP Journal on Wireless Communications and Networking is to bring together science and applications of wireless communications and networking technologies with emphasis on signal processing techniques and tools. It is directed at both practicing engineers and academic researchers. EURASIP Journal on Wireless Communications and Networking will highlight the continued growth and new challenges in wireless technology, for both application development and basic research. Papers should emphasize original results relating to the theory and/or applications of wireless communications and networking. Review articles, especially those emphasizing multidisciplinary views of communications and networking, are also welcome.

http://jwcn.eurasipjournals.springeropen.com/

14. Mobile Networks and Applications

The journal Mobile Networks and Applications reflects the emerging symbiosis of portable computers and wireless networks, addressing the convergence of mobility, computing and information organization, access and management. In its special issues, the journal places an equal emphasis on various areas of nomadic computing, data management, related software and hardware technologies, and mobile user services, alongside more "classical" topics in wireless and mobile networking. The journal documents practical and theoretical results which make a fundamental contribution.

http://www.springer.com/engineering/signals/journal/11036/PSE

Unit 10

IoT for Smart Grid

Passage A   The Introduction of Smart Grid

Passage B   An IoT-based Energy-management Platform for Industrial Facilities

Passage C   Smart Grid Applications

Passage A　The Introduction of Smart Grid

Ⅰ. How Can We Define the Smart Grid

There is no single definition of the Smart Grid. Instead, there is a set of expectations that must be met to face the wide range of new requirements exposed in the area. The Smart Grid must enhance the current grid network with advanced sensing actuators and a highly secure networking infrastructure to improve grid efficiency, performance, and reliability as well as to support a wide range of new services (e.g., better knowledge of power consumption profiles, use of PHEV, distributed sources such as solar panel and residential power generation, and smart home appliances).

The Smart Grid is one of the major applications for smart object networks, and the IP protocol will be central to them. As described through several use cases explored in this chapter, most of the expectations and requirements for the Smart Grid involve smart object networks: sensors (e.g., measuring the current, voltage, phase, or reactive power) and actuators (e.g., circuit breakers, etc.) to efficiently monitor and control the power grid, sensing in smart meters to measure power consumption, and a number of smart devices used in homes, buildings, and factories that communicate via specialized energy management devices with the grid for efficient energy management.

A typical power grid architecture from power generation to the home/building is depicted in Figure 10.1: power is generated by plants and then distributed to the end user through a distribution network. The particularity of the grid network lies in its hierarchical structure. High voltage (HV) lines are connected to (primary) substations where the voltage is reduced to MV before being even further reduced to LV using pole tops (United States) or secondary substations (Europe). Finally, electricity is delivered to the end user where a smart meter is used to monitor energy usage (and to perform many other functions).

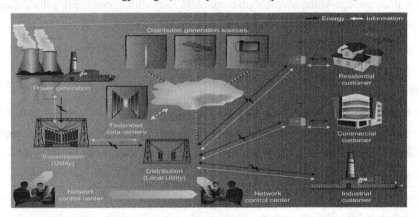

Figure 10.1　Typical power grid architecture

Ⅱ. Features of Smart Grid

The smart grid represents the full suite of current and proposed responses to the challenges of electricity supply. Because of the diverse range of factors there are numerous competing taxonomies and no agreement on a universal definition. Nevertheless, one possible categorization is given here.

1. Reliability

The smart grid will make use of technologies, such as state estimation, that improve fault detection and allow self-healing of the network without the intervention of technicians. This will ensure more reliable supply of electricity, and reduced vulnerability to natural disasters or attack.

Although multiple routes are touted as a feature of the smart grid, the old grid also featured multiple routes. Initial power lines in the grid were built using a radial model, later connectivity was guaranteed via multiple routes, referred to as a network structure. However, this created a new problem: if the current flow or related effects across the network exceed the limits of any particular network element, it could fail, and the current would be shunted to other network elements, which eventually may fail also, causing a domino effect. See power outage. A technique to prevent this is load shedding by rolling blackout or voltage reduction (brownout).

2. Flexibility in network Topology

Next-generation transmission and distribution infrastructure will be better able to handle possible bidirection energy flows, allowing for distributed generation such as from photovoltaic panels on building roofs, but also the use of fuel cells, charging to/from the batteries of electric cars, wind turbines, pumped hydroelectric power, and other sources.

Classic grids were designed for one-way flow of electricity, but if a local sub-network generates more power than it is consuming, the reverse flow can raise safety and reliability issues. A smart grid aims to manage these situations.

3. Efficiency

Numerous contributions to overall improvement of the efficiency of energy infrastructure are anticipated from the deployment of smart grid technology, in particular including demand-side management, for example turning off air conditioners during short-term spikes in electricity price, reducing the voltage when possible on distribution lines through Voltage/VAR Optimization (VVO), eliminating truck-rolls for meter reading, and reducing truck-rolls by improved outage management using data from Advanced Metering Infrastructure systems. The overall effect is less redundancy in transmission and distribution lines, and greater utilization of generators, leading to lower power prices.

4. Sustainability

The improved flexibility of the smart grid permits greater penetration of highly variable renewable energy sources such as solar power and wind power, even without the addition of energy storage. Current network infrastructure is not built to allow for many distributed feed-in points, and typically even if some feed-in is allowed at the local (distribution) level, the transmission-level infrastructure cannot accommodate it. Rapid fluctuations in distributed generation, such as due to cloudy or gusty weather, present significant challenges to power engineers who need to ensure stable power levels through varying the output of the more controllable generators such as gas turbines and hydroelectric generators. Smart grid technology is a necessary condition for very large amounts of renewable electricity on the grid for this reason.

III. Why Is Smart Grid is Important

Global electrical grids are verging on the largest technological transformation since the introduction of electricity to the home. Current day electrical power infrastructure, which has served us well without major modernization, is rapidly running up against its limitations. Energy production is shifting from centralized to distributed creating bi-directional power flows forcing an alteration from a radial to mesh topology. The Department of Resources, Energy & Transportation (Australia) has ultimately concluded that the existing grid infrastructure requires upgrades and redesign to facilitate this shift from centralized to distributed power generation.

The current power grid must be expanded in order to provide support for the expected increase in distributed generation and specifically look at both local and system-wide impacts of integration of distributed generation, including a high percentage of renewables. Due to the variability and limited predictability of the weather, renewable power sources can produce voltage fluctuations and the occurrence of harmonics in distribution lines. It is therefore apparent that a robust, high speed, bi-directional communications network is required to provide a higher resolution picture of the Smart Grid. This network will overlay the current power grid and provide unprecedented control and monitoring functions, whilst also driving autonomous functions and utilizing smart algorithms.

It is unlikely that grid operators will be willing to build whole telecommunications networks to control and monitor the Smart Grid instead opting to use a privately owned wireless access network using current technology such as Long Term Evolution (LTE). However, this increases the communication traffic in an environment where communication traffic is already high. By optimally allocating communication resources, this research shows that the effect of adding the Smart Grid data class can be minimized and QoS properties retained.

IV. What Is Smart Metering

The smart meter will provide the means of communication between consumers and the utilities. This will enable the integration of other technologies such as demand response. Real-time consumption level data can be transferred to the utilities, and will enable the consumers to monitor their electricity consumption and take measures to reduce their usage. Additionally, smart meters can provide real time pricing of the electricity or indirect load control known as dynamic price response including:

1. Time of Use (ToU) Tariff

This scheme encourages consumers to shift their consumption from a peak to off peak period. This is a non-dynamic tariff, and in fact large scale integration of such a tariff may in turn just simply shift the peak time.

2. Real-Time Pricing (RTP)

The price of the electricity in the market changes hourly (or half an hourly in some markets). The RTP programmers offer a type of tariff which changes hourly to reflect the variations in supply and demand and the price of the electricity in the market. It provides incentives to consumers to limit their consumption when the wholesale price of the electricity is high and increase their consumption at lower electricity price periods.

3. Coincident Peak Pricing (COP) or Known as Critical Peak Pricing (CPP)

RTP is deemed infeasible for residential customers, a reasonable alternative is critical-peak pricing (CPP). CPP tariffs augment a time-invariant or TOU rate structure with a dispatchable high or "critical" price during periods of system stress. The critical price can occur for a limited number of discretionary days per year, or when system or market conditions meet pre-defined conditions. Participating customers receive notification of the dispatchable high price, typically a day in advance, and in some cases are provided with automated control technologies to support efficient load drop. Because all of the prices in a CPP rate are preset, CPP is not as economically efficient as RTP; this same characteristic, however, also makes CPP politically more appealing, because it diminishes the potentially large price risk associated with RTP.

Another major benefit of smart meters is detection of electricity theft which account of some of the revenue of utilities. Since the communication is happening at real-time, any tampering with the metering equipment, or bypassing the meter can be transmitted to the utility.

V. The Current Communication Standards of Smart Metering

Work on standards for smart metering communications has been underway for many years however due to the recent pace of technological advancements in several communication sectors the bandwidth available for data transmission has risen exponentially.

Current communication standards for Smart Metering can be distinguished by two main communication mediums, namely wired and wireless. They are however intertwined with some systems that employ a hybrid technique to merge the advantages of low-power RF communication with PLC.

1. Wired Communication

Power Line Communication (PLC): The most dominant technology in wired smart metering communications is Power Line Communication (PLC), also known as Power Line Carrier. PLC has evolved since 1980 and is an obvious first choice for communicating with electricity meter as a direct connection to the meter is already installed. However due to the communication signal being directly imposed onto the power line there is a technical issue with radiation of the signal into the environment surrounding the power lines. High bandwidth PLC has been shown to interfere with the DGPS band used for geo-positioning.

Bandwidth requirements also highlight the communication medium as insufficient due to the lack of distinct ohmic separated circuits available. With most transmission lines being three phase and communications limited to low frequency bands, the potential total bandwidth of a channel between two points is limited.

Efforts towards investigating the influence of time-varying channels on PLC equipment was focused towards frequency response in the tens of MHz. However, attenuation of these frequencies over long lengths of transmission line poses problems for channel crosstalk. In order to ensure the robustness of the communication channels an element of channel coding must be employed. By using linear block codes to select a suitable frequency to communicate between two nodes coded OFDM can operate in channels with a decreased Signal to Noise Ratio SNR.

Further optimization of the PLC channel can be derived from in-depth analysis of the multipath behavior of interconnected nodes within the power distribution system. The attenuation effects upon the signal propagating from Low Voltage substations to customer premises can be shown to include delay effects that are inherently cyclic in nature leading to inefficiencies in any phase-based coding scheme, especially due to the non-linear delay function.

2. Wireless Communication

Wireless communications represent themselves as an alternative to using wired technologies, advantages over wired implementations are mainly placed within the reduced cost of infrastructure as no physical cabling is required between the separated nodes in the wireless system. There are however concerns over interception of the data between nodes as it is propagated through physical space and can be picked up by any person within range of the transmitting node.

(1) Low-Power Wireless: The need for low-power transceivers is apparent when the application of measurement technologies is reviewed. If the impetus to begin real-time monitoring of nodes consuming or producing power within the transmission network is driven to its logical conclusion, there is a possible need to monitor each appliance connected to the in-home power distribution network to enhance the ability to measure, monitor and control.

(2) ZigBee (IEEE 802.15.4): ZigBee represents itself as the communications standard of choice for low power operation, although concerns have been raised towards the robustness of ZigBee under noise conditions. The interference of 802.11/b/g within short distances of ZigBee nodes can obliterate the entire communications channel rendering it completely unusable and impractical for reliable "always-on" connections.

(3) Z-Wave: Z-Wave, a closed source, proprietary alternative to ZigBee claims to have better robustness to interference from 802.11/b/g due to its operating within the 800MHz range, well below 802.11/b/g.

(4) Mesh Networks: Mesh Networking protocols eradicate the need for each node in a system to be within transmission range of a master node that is the endpoint for communication. Instead packets of information are able to be exchanged with neighboring nodes that can then subsequently pass on the information towards their neighbors until the packet reaches its intended recipient.

This passing of packets between nodes allows large distances between a source and destination of data; however, in a network with many connected nodes, a packet passed between neighbors could loop around and be transferred back to nodes surrounding the originator. This inevitably causes an overhead in the communications channel that can reduce the effective bandwidth available for data payload.

VI. The Application of the Smart Metering

At present, in some European countries old electricity meters have been replaced by smart meters. In Italy there is a large distributor (ENEL) and many medium-sized distributors owned by Municipalities (Rome, Milan, Turin, Brescia, Parma, Verona, Trieste, Bologna, etc.), and of smaller local distributors. An automatic (Smart) Metering scheme is mandatory for ENEL, while the rest of the distributors continue to use the old metering systems. Generally, meters are owned by distributors and customers are not allowed to buy their own meters. ENEL has managed to install nearly 27 million meters over five years from 2002, with an investment over 2.2 billion Euro. Different types of customer have been given smart meters, and as December 2006 over 95% of the electricity meters owned by ENEL were smart meters.

ERDF, a subsidiary of EDF, and the largest electricity distribution network in the European Union; deals with 33 million customers in France. EDF announced the first phase of a nationwide rollout of 35 million smart meters in July 2008. The consortium, led by Atos Origin, the international information technology company, will conduct a pilot project encompassing 300,000 Power Line Communication (PLC) meters and 7,000 data concentrators in France. The project roll-out is expected to start early in 2010.

SCE had several pilots for automated meter reading (AMR) in the 1990s. While these pilots did not lead to full-scale implementations at the time, they were valuable for identifying important requirements for later pilots and finally the implementation of a fully automated AMI system. In the 2000s, the California Public Utilities Commission (CPUC) asked the utilities to investigate the use of AMI/AMR technology. As a result of this request, SCE initiated its Smart Connect program. This program intends on installing smart meters for all of its customers and integrating them with the utility's customer conservation and incentive programs.

Ⅶ. New Words

Bandwidth ['bændwɪdθ]　　　　　　n. 带宽；频带宽度
Frequency ['fri:kw(ə)nsɪ]　　　　　n. 频率；频繁

Ⅷ. Phrases

Smart Grid	智能电网
solar panel	太阳电池板
residential power generation	住宅发电
smart home	智能家居
smart object networks	智能对象网络
power grid	电力网
smart meters	智能表
hierarchical structure	分级结构
radial model	射线模型
voltage reduction	电压下降
bidirection energy flows	双向能量流动
hydroelectric power	水力发电
local sub-network	本地子网
Advanced Metering Infrastructure systems	先进的计量基础设施系统
bi-directional communications network	双向通信网络
Signal to Noise Ratio	信噪比
Low-Power Wireless	低功耗无线
Mesh Networking protocols	网状网络协议

Ⅸ. Abbreviations

HV	High Voltage	高电压
VVO	Voltage/VAR Optimization	电压/无功优化
LTE	Long Term Evolution	长期演进
COP	Coincident Peak Pricing	尖峰分时电价
CPP	Critical Peak Pricing	尖峰电价
PLC	Power Line Communication	电力线载波通信
AMR	automated meter reading	自动读表

Passage B An IoT-based Energy-management Platform for Industrial Facilities

Ⅰ. Introduction

Interconnectivity and interoperability are very important features in the development of integrated energy management systems for industrial facilities. A simple and common strategy for exchanging energy-related information among the entities in a facility is currently lacking. To this end, the purpose of this study is to present an IoT-based communication framework with a common information model to facilitate the development of a demand response (DR) energy management system for industrial customers. Additionally, we developed and implemented an IoT-based energy-management platform based on a common information model and open communication protocols, which takes advantage of integrated energy supply networks to deploy DR energy management in an industrial facility. The experimental results of this study demonstrate that the proposed platform can not only improve the interconnectivity of the entities in industrial energy management systems but also reduce the energy costs of industrial facilities.

Industry is the largest consumer of electricity among all end-user sectors. According to statistics from the International Energy Agency, in 2012, the consumption of electricity worldwide by the industrial sector was 42.3% of total energy produced. This has led to significant interest in the development of industrial energy management around the world in recent years.

The variety of energy management systems, integrated energy systems (IESs) and customer equipment can generate many different data formats, resulting in a lack of interconnectivity and interoperability in industrial energy management systems. The main goal in the integration of energy management systems is to allow the communicating entities to interact with each other using a common information model. This results in the need to be able to understand and map the meaning and context of information that resides in different domains. In view of this requirement, a common information model is needed to support the ability to semantically align the information for inter-domain communication for the energy management systems in industrial facilities.

The concept of a metamodel for information integration is not new, and recently the use of the Extensible Markup Language (XML) as a metadata exchange technology has been widely adopted for providing better information integration between component systems. The requirements of semantic interoperability are now being met on the customer side by the use of the Facility Smart Grid Information Model (FSGIM). The FSGIM is a specification that is currently under development by the American Society of Heating and Air-Conditioning Engineers (ASHRAE). It provides a common data structure that can be used for developing energy management systems in different facility environments. However, currently it does not take into account the specific use of FSGIM in industrial processes. To apply the FSGIM to an industrial domain, some extensions and attributes need to be added. In this way, the common information model could be extended to be adaptable for various communication protocols that have already been applied within industrial facilities.

Many industrial communication protocols can be adopted to support FSGIM for industrial energy management. However, applying different private protocols will result in poor interoperability and high development cost. An alternative is the Internet of Things (IoT), an infrastructure of interconnected objects, people, and systems, together with information resources and intelligent services. By using the IoT to interconnect the devices, it is expected to make industrial production more intelligent and efficient. Moreover, energy management systems for industrial facilities would no longer be stand-alone entities but part of ubiquitous networks. IESs in industrial facilities can be scheduled efficiently and used to their full capacity.

From the viewpoint of energy management schemes, a demand response (DR) program is a key technology that induces users to modify their consumption patterns in response to electricity prices or payment incentives. From the customer side, one of the main purposes of deploying DR is to reduce electricity costs by using more power during times of low prices and less power during times of high prices. Use of DR has been widely studied in commercial and residential facilities, but rarely in industrial. Because manufacturers need to consider not only the overall usage of electricity but also specific resource demands during the operation and production processes, it is difficult to design and implement an IoT-based DR energy management system for industrial consumers. In this study, we focus on designing a common information model and the related IoT-based communication framework for DR management in industrial facilities.

Because of the high electricity consumption in industrial applications, it is imperative to establish an industry-centric energy management system as well as to make sure that this system is utilized in a proper way. A common information model plays a significant role in achieving the interoperability of energy networks between smart grids and facilities. As defined by the National Institute of Standards and Technology (NIST), the utility meter and the energy service interface (ESI) are placed at the boundary of a facility and exchange communication data between the customer domain and other external domains (e.g., distribution, operation, and market). The other domains in the smart grid provide external energy services to the customer domain. In general, to cope with heterogeneity, a smart grid adopts a canonical data model (CDM) approach. The conceptual model of an industrial facility with a smart grid is shown in Figure 10.2. At the grid side, IEC common information model (CIM) families of standards act as the main information model standards. As shown in Figure 10.2, open automated demand response (OpenADR) specifications can be considered as the smart grid user interface bridge between the grid and the facility. The common information standard used at industrial facilities is the FSGIM, which consists of Load, Meter, EM (Energy Manager) and Generator components.

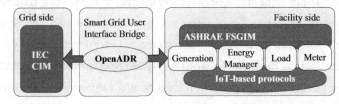

Figure 10.2　Smart grid information model standards and relationships between standards

IoT-based protocols can be used within data acquisition and control systems to sense, gather, store, analyze, display and control internal facility processes. The important factor is that industrial IoT-based technology should be fully compatible with IP on the communication stack, which provides the possibility to enhance energy management efficiency. There are some studies that have proposed an IoT-based cloud manufacturing service system and its architecture. With an IoT-based framework, the number of networked sensors grows dramatically across production, supply chains and products. Shrouf presented a reference architecture for IoT-based smart factories and proposed an approach for energy management based on the IoT paradigm.

To enhance energy management efficiency, a great deal of effort has been focused on DR energy management implementations. Previous studies proposed a DR energy management model and algorithm for industrial facilities based on a state-task network (STN). However, most of the existing work focused on DR energy management without considering the integration and interoperation of different systems. To achieve a wide scale DR application, more abstract information models and standardized communication protocols need to be considered. To this end, we propose a common information model and IoT-based communication framework that can enhance the interconnectivity and interoperability of energy management systems.

II. The Proposed Information Model and Communication Framework

Until now, the CIM has been used only to provide interoperability at the grid side. However, the FSGIM is a currently available facility-side specification. The FSGIM defines an abstract, object-oriented information model to enable control systems to manage electrical loads and generation sources in response to communications with an electrical grid. In this paper, the FSGIM is utilized to represent the energy consumption, production, and storage systems in an industrial facility. As a result, a level of interoperability is ensured because the information has been standardized.

Furthermore, the FSGIM defines the data elements, data type, data associations, semantic checks and data optionality. The FSGIM can be used to develop or enhance other standards that define technology and communication protocol specific implementations. It provides the basis for interoperable extensions to the existing communication protocols for facility information. Figure 10.3 illustrates the types of information that are standardized in the FSGIM and the relationship between the FSGIM and the communication protocol.

As shown in the middle part of Fig. 10.3, when applying the FSGIM to industrial energy management systems, development is needed for the communication protocols at different layers, as well as security and additional services. These protocols may use their own existing mechanisms to perform the information encoding and communication. As shown on the right side of Figure 10.3, there are several IoT-based communication protocols and solutions that can be used to support FSGIM in industrial energy management systems. IoT-based protocols have several advantages, including an effectively limitless ability to scale, and the ability to accommodate multiple millions of end nodes on a single network.

The purpose of using an IoT-based energy management framework is to allow the system to achieve IP-based remote access through open protocols. Using IoT promotes the ability to exchange energy-related data (collected from ubiquitous devices on the plant floor and from the enterprise energy manager), which is expected to increase cost savings. An IoT communication service with enhanced communication and sensing capabilities enables interactions between energy producers and end users, and also facilitates the control of distributed energy resources (DERs) and their integration into the main grid.

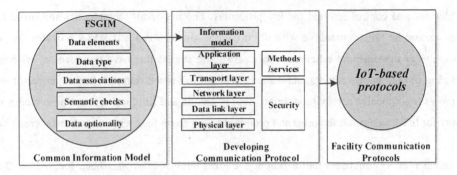

Figure 10.3 The relationship between FSGIM and communication protocols

To implement the FSGIM in industrial facilities, the protocols for both the backbone and wireless networks need to be designed. In this paper, we propose the use of the following communication protocols:

1. Physical and Data Link Layer

In this study, we propose using Industrial Ethernet technology, because it has been widely deployed and has promoted the convergence of plant control and enterprise networks. Additionally, Industrial Ethernet's simple and effective design has made it the most popular networking solution at the physical and data-link levels. Examples of Industrial Ethernet include PROFINET, EtherCAT, Ethernet Powerlink, RAPIEnet and EPA. Energy decision-makers can take advantage of real-time information by accessing key performance indicators and data analytics at the manufacturing application level. The industrial process can be monitored and adjusted in real-time to improve production flexibility.

For the field network, IEEE 802.15.4 is used, which is a standard radio technology for low-power, low-data-rate applications. Because of the ubiquity and the availability of IEEE 802.15.4-compliant radio transceivers, many of the recently developed industrial radio stacks are built on IEEE 802.15.4, including ISA100.11a, WirelessHART and WIA-PA.

2. Network Layer

The opportunity to use open protocols such as IP over Ethernet networks offers the possibility of a level of standardization and interoperability in the industrial field. The transition of IP networks from IPv4 to IPv6 is ongoing. New or improved features, such as transparent end-to-end communication, large addressing space, an auto-addressing method, a more efficient routing protocol, enhanced mobile capability, and autonomous network forming and configuration are very attractive and useful for a plant control network. Plant control networks will migrate to IPv6 to integrate with enterprise intranets and the Internet in the near future.

Besides the IPv6 transition and the integration of different networks, the emergence of IP-enabled wireless field networks is another important trend. The use of IP-based wireless technologies in industrial energy management provides new possibilities and advantages compared to existing wired solutions. These technologies will enable easier access to more information that is related to the process itself and the equipment used in the process.

Lightweight IP stacks and the IPv6-based communication protocol make it possible to enable IP communication in wireless field networks. 6LoWPAN is an adapter layer between the IPv6 and IEEE 802.15.4. It is used for low-power and lossy networks (LLNs) where the links used for interconnecting the nodes are IEEE 802.15.4 links. The 6LoWPAN can even be applied to very small devices, including low-power devices with limited processing capabilities allowing them to participate in the IoT. In this work, the energy manager, the load, and the generation systems are located in an industrial facility and are connected by either a wireless or a wired network. The particular communication technology used in the lower layers can be disregarded because the system is based on IP at the network layer.

3. Transport Layer

TCP or UDP can be chosen for IoT-based backbone and wireless networks. UDP over 6LoWPAN fits well with many smart object applications. UDP provides a best-effort datagram delivery service, but does not guarantee that the datagrams are delivered to the destination. It is up to the application layer to recover from packet loss. However, the simplicity and lightweight nature of UDP makes it a compelling choice for data that need to be transported quickly such as sensor data.

4. Application Layer

For industrial wired part, we chose application layer protocols such as the Hypertext Transfer Protocol (HTTP), for web-style interaction and web service infrastructure, and the Simple Network Management Protocol (SNMP) for network configuration. These allow IP-based smart objects to interoperate with a large number of external systems.

For the wireless part, the application layer protocol also needed to be considered. Constrained Application Protocol (CoAP) is an application layer protocol that is intended for use in resource-constrained nodes. It is particularly targeted at small low-power sensors, switches, valves and similar components that need to be controlled or supervised remotely through standard Internet networks. CoAP is designed as an easy translation of HTTP for simplified integration with the web, and also meets specialized requirements such as low overhead and simplicity. This makes it possible to use CoAP in one-to-many and many-to-one communication patterns.

In this study, CoAP is used as the application layer protocol for implementing the DR scheme in the wireless part.

5. Web Service System

By using web service technology for industrial energy applications, existing web-service-oriented systems, programming libraries, and knowledge can be directly applied. For energy networks, smart energy supply networks and applications can be directly integrated with existing management systems by using the same interfaces. This makes it possible to integrate energy applications into enterprise resource planning systems without any intermediaries, thus reducing the complexity of the system as a whole. For industries, energy management applications can be built using off-the-shelf technology without any customized interfaces or translators.

6. Service Discovery

The IP architecture does not have any default service discovery framework. Among the service discovery mechanisms, auto address configuration is especially important, which can be done either with a centralized protocol such as the Dynamic Host Configuration Protocol (DHCP) or with a distributed mechanism such as IPv4 auto address configuration or IPv6 stateless address configuration. Additionally, some alternative mechanisms such as Service Location Protocol (SLP), Zeroconf and Universal Plug and Play (UPnP) can be used in an industrial energy management system.

7. Security

Security consists of three properties: confidentiality, integrity, and availability. To implement security architecture, encryption can be adopted to convert messages from plaintext into ciphertext, which is not readable by potential attackers. Several other security mechanisms can also be used for energy management systems, such as authentication and key distribution. For computationally constrained smart object microprocessors, hardware-assisted encryption implementations can be adopted to enable strong encryption.

III. IoT-Based DR Energy Management System with FSGIM

This section discusses the concept of an IoT-based industrial energy-management platform for DR in industrial facilities. The architecture and the interrelationship among all the model elements are shown in figure 10.4.

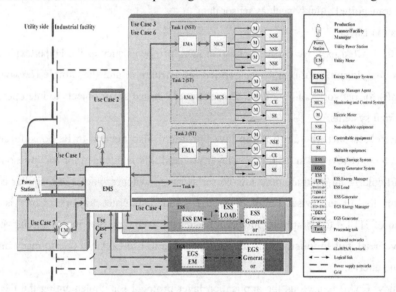

Figure 10.4 System architecture

1. System Architecture

The proposed system architecture is divided into the utility side and industrial electricity demand side, with the utility meter as the boundary between the two. As described below, the energy supply networks, i.e., utility power station, energy storage system and energy generation system are integrated with the process tasks as an integrated system.

System elements are essential for building the automatic DR energy-management platform. Each element is uniquely identified by a single symbol, as illustrated in the right box of Figure 10.4.

(1) Production Planner/Facility Manager is responsible for carrying out production plans. It responds to real-time changes based on feed-back from a process as well as other internal or external events, and is responsible for the maintenance and operation of the facility.

(2) Utility Power Station acts as an energy supplier and energy information provider. It interacts with the EMS and the smart grid owned Smart Meter.

(3) Utility Meter measures energy consumption or generation per time tariff and provides this information to the utility company. Secure communications with this device are provided by the smart grid.

(4) Energy Manager System (EMS) is any device/software or group of the two, installed in an industrial

facility that provides the functions of energy management, control and planning in conjunction with responsible facility management. EMS runs the DR algorithm to determine the optimal operating points of tasks and the operating status of IESs to shift the electricity demand from peak to off-peak demand periods, and transmits the control information to MCS. The EMS is located at a top hierarchical level of the industrial network, and can manage and monitor task-level attributes and energy costs of all tasks.

(5) Energy Manager Agent (EMA) monitors the electricity consumption and controls the electric load of each task. The EMA is located at a lower hierarchical level than the EMS in the industrial network, and can manage task-level attributes and energy costs of specific task.

(6) Monitoring and Control System (MCS) is an automation system designed to monitor and control the operation of each device. The MCS is located at the same hierarchical level as the EMA in the industrial network, which can control and monitor load-level attributes and energy costs of specific task.

(7) Meter (M) is a physical device or subsystem onto which an electric meter is defined.

(8) Non-shiftable Equipment (NSE) is a device whose energy demand must be met immediately.

(9) Controllable Equipment (CE) is a device that has multiple operating levels, resulting in differences in electricity demand.

(10) Shiftable Equipment (SE) is a device that can be switched on or off based on the electricity demand.

Energy Storage System (ESS) is a physical device or subsystem that can store electrical energy and whose electrical energy can then be delivered at a later time. For example, a rechargeable battery. ESS is logically divided into ESS EM, ESS Load and ESS Generator.

(11) ESS EM acts as an energy manager performing an internal energy management function to control the electrical storage device.

(12) ESS Load acts as a load while the storage device is being charged.

(13) ESS Generator acts as a generator while the storage device is supplying electrical energy.

(14) Electricity Generation System (EGS) generates electricity in an industrial facility, including solar panels, wind turbines, waste heat recovery, etc. EGS is logically divided into EGS EM and EGS Generator.

(15) EGS EM performs an internal energy management function to control the electricity generation system.

(16) EGS Generator acts as a generator while the electricity generation system is supplying electrical energy.

The blue box represents the process task in industrial applications. The red line represents the electric grid and the red dashed line represents the power supply networks inside the industrial facility. The bold blue arrow line represents the IP-based backbone network and the black arrow line represents 6LoWPAN networks, which were discussed in Section 4. The dashed black arrow line represents the logical links inside ESS and EGS.

2. Common Industrial Information Model

This section discusses the information models and their instances for this system. As mentioned above, the metadata model in FSGIM XML is a representation of the energy management system proprietary data model and based on open standards that support interoperability. In this study, we consolidate these different metadata representations into a semantically aligned representation of the network reality.

In industrial applications, it is difficult to use the FSGIM to represent all network data parameters. We need to exploit data from industrial applications not previously modeled in the FSGIM, and then semantic alignment is available through harmonization. To do this, we added some additional parameters. This is possible at the metamodel level of abstraction and is referenced in the FSGIM semantic models.

The information model between Power Station and EMS is based on an Energy Market Information Exchange (EMIX), which defines price representations and market interactions. The relationship between the information model and its instance in DR energy management for an industrial facility is shown in Figure 10.5. The main common information models that we used in this study include the load model, the meter model, the EM model and the generator model. In Figure 10.5, the yellow box represents UM and meter, which are instances of the meter model. The pink box represents equipment, which are instances of the load model. The brown box stands for the energy supply parties, which are instances of the generator model. The green box represents the EM, which are instance of the EM model. We will discuss them as follows:

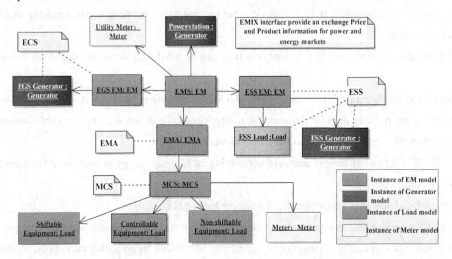

Figure 10.5   The relationship between the information models and their instances in DR energy management for industrial facilities

(1) Load model

The *Load* component defines the common attributes to all devices that consume electricity. In the DR model, processing tasks are divided into non-schedulable tasks (NSTs) and schedulable tasks (STs), and both are composed of pieces of industrial equipment. NSTs are tasks for which the demand cannot be scheduled, and must be satisfied immediately despite the fact that the price of electricity is high or low. STs are tasks for which the demand can be scheduled among a pre-specified set of operating points.

The industrial equipment is classified as NSEs, SEs and CEs. NSTs must contain NSEs only, and STs consist of NESs, SEs and CEs. The FSGIM *Load* component is used directly to represent NSEs, SEs and CEs.

In the FSGIM, *Curtailable Load* is a child class of Load. The *Curtailable Load* class defines attributes that are unique to devices whose electrical consumption can be curtailed. The attributes define the information that is needed to represent curtailment policies based on criteria including load priorities, cost constraints, protection of physical equipment, and safety considerations. The *Curtailment Ratings Level* is an attribute of *Curtailable Load*. This class defines a single array. Each array element defines the fixed demand to which the load shall be controlled as a function of the stated level. The *Priority* is an attribute of *Curtailable Load*, which defines the curtailment priority and the order upon which curtailment occurs across a set of loads. A *Priority* value of 1 is considered a critical load that may not be shed except during a critical event.

- CE is modeled as an instance of the *Curtailable Load* class model.
- SE is modeled as an instance of the *Curtailable Load* class with one curtailment ratings level.
- NSE is modeled as an instance of the *Curtailable Load* class model with priority equal to 1, i.e., the highest priority.

(2) Meter Model

A meter is modeled as an instance of the *Meter* class in the FSGIM. We use the FSGIM *Meter* component directly, which provides an abstract representation for the function of measuring power, energy, and emissions. The instance of meter is a physical device or subsystem which measures the demand or supply of one or more loads and/or one or more generators.

(3) Energy Manager (EM) model

The FSGIM defines the *EM* component to provide an abstract representation of energy management system functionality. The functionality is implemented in any device that performs analysis and makes energy-related decisions involving meters, loads, generators, or energy storage devices. In this platform, EMS, EMA and MCS are instances of the *EM* model.

In view of this situation, it is difficult to use the *EM* model that is defined in the FSGIM to represent the industrial STN elements completely, which allows for development of a general algorithm to find an optimal solution for the factory energy management system. The EMS, EMA and MCS models need to be designed by adding some classes based on the original FSGIM *EM* model.

EMS model: The EMS model includes the necessary attributes for managing and monitoring the electricity demand of whole industrial facilities. As a result, it can maintain the task attributes and operation information of whole industrial applications. The EMS as the instance of the EMS model establishes communications with the utility suppliers via a wide area network (WAN) to obtain the day-ahead dynamic electricity prices. The EMS transmits control commands to each MCS via the industrial Ethernet backbone network.

- EMA model: The EMA model includes the necessary attributes for managing and monitoring energy for a task. As a result, it can maintain the task attributes and operation information of the specific task.
- MCS model: The MCS model includes the necessary attributes for controlling and measuring the energy consumption of each processing task. As a result, it maintains the task attributes and load operation point attributes of a specific task. It controls the field devices to operate at specific operating points requested by the EMS by sending control commands

(4) Generator model

We use the FSGIM *Generator* component directly. Many classes of the *Generator* component are derived from the IEC 61850-7-420 standard. The *Generator* component provides an abstract representation of all devices that produce electricity or store it. In our platform, the generation element of EGS and ESS can be considered as an instance of the *Generator* model.

The ESS Generator is an instance of the *Generator* model while the storage device is providing electrical energy. The EGS Generator is also an instance of the *Generator* model while the EGS is supplying electrical energy. A power station can also be considered as an instance of the *Generator* model.

3. Use Cases

Several DR-related case studies have been conducted to examine the availability of the IoT-based smart grid platform with a common information model for industrial facilities. The use cases are presented in this section.

(1) Use Case 1: Determining energy/demand price information

In this use case, the power station provides the dynamic pricing to the facility. These price data are developed by the power station using internal procedures to maintain a balance between generation and supply near the time of use. Typically these price data are used by facilities to manage their current operating consumption. The price data are used by the facility control systems to control resources, subject to ensuring that all production safety, performance and product quality requirements are maintained. The communication between the utility power station and the EMS is based on using a wide area network (WAN), in which TCP/IP can guarantee reliability of packet transfer.

(2) Use Case 2: Determining DR parameters

In this use case, the EMS prepares the parameters of the DR algorithm and makes the DR decision. The DR algorithm is formulated using the STN model and mixed integer linear programming (MILP). These parameters include the STN model parameters, the supported operating point information of each task, the operating parameters of the EGS and the operating parameters of the ESS.

Based on the inputs including the day-head electricity price, the STN representation of industrial facilities, the operating information of each task, and the operating information of DERs, the DR algorithm selects the optimal operating point of STs for each pre-specified time interval. By scheduling STs with different operating points, the DR algorithm can shift part of the electricity demand from peak to off-peak demand periods.

The DR algorithm also determines the optimal power source (i.e., the grid, EGSs, or ESSs) for each time interval. For example, during off-peak periods, the electrical grid supplies electricity to industrial facilities, and the ESS charges energy from the power grid, whereas during peak periods, the EGS (which may be solar, wind, or waste heat from power plants) supplies electricity, and the ESS discharges to supply energy to industrial facilities. The detailed DR algorithm for industrial facilities is described in.

(3) Use Case 3: Managing the operation point of each time interval to minimize cost

In this use case, the facility manages the operating point of each time interval to minimize cost. After making a DR decision, the EMS provides the EMA of each task, the optimal operating point of the task and the operating time of that operating point. The EMA proposes operating levels of the equipment in each task and sends this information to the MCS, which finally controls the equipment according to the commands scheduled by the EMA. When the state of the load changes after receiving a control message, the wireless equipment relays the updated state to the MCS and EMA. The EMA sends the task level response to the EMS. These response messages not only act as the acknowledgement but also carry the important attributes.

(4) Use Case 4: Determining the utilization of ESS

In this use case, the facility decides to buy power or utilize the ESS within each time interval. After determining the energy price information and the DR parameters, the EMS specifies the ESS operating mode in the next stage and the operating duration of that mode, which is an IES decision. The communication between the EMS and ESS is based on TCP, which guarantees reliability of packet transfers.

(5) Use Case 5: Determining the utilization EGS

In this use case, the facility decides to buy power from the grid or use the EGS within each time interval. After making a DR decision, the EMS specifies the EGS operating mode in the next stage and the operating duration of that mode, which is an IES decision. In a high-electricity-price period, it encourages the EMS to command the EGS to supply electricity to some or all of the processing tasks. This decreases the electricity demand of the industrial facility.

(6) Use Case 6: Measuring equipment power consumption

In this use case, the facility measures equipment power consumption for each task and each load. The facility measures the power consumption of a particular electrical device and each task. Prior to its use, the operation manager installs a meter to measure energy at a device or for each task. Periodically, the meter measures the energy consumption of each device and sends an energy measurement message to the MCS to track the ongoing energy use of the load. The MCS sends the energy consumption of related loads to the EMA. The EMA sends the energy consumption of each task to the EMS. The EMS supplies the energy consumption to the facility manager.

(7) Use Case 7: Measuring all energy consumption in a factory

In this use case, the facility measures all equipment power and provides this information to the utility power station and the EMS. The utility meter measures the whole energy consumption of the factory. The utility meter sends the energy measurement message to the EMS to track the ongoing energy use of the factory. The EMS provides the energy consumption information to the facility manager. The Utility Meter sends the energy measurement message to the Utility Power Station, which is used by an energy provider to track facility or equipment performance and analysis.

IV. Development of the Experimental Platform

In this section, we introduce the integrated hardware and software experimental platform that we developed in this study. The solution was based on private protocols with high expenses and the data structure was not based on a common information model; as a result the interactions and interconnections for the energy management networks were extremely complex. In this study, a new industrial energy-management platform has been prototyped and developed. We implemented IP-based Ethernet as the backbone network. Meanwhile, we implemented 6LoWPAN over IEEE 802.15.4 as the wireless part. Here, the field node applications can be directly integrated with existing IP-based systems and use the standardized interfaces, which drastically reduces the complexity of the system as a whole. Moreover, we implemented HTTP for web-style interactions in the wired part and implemented CoAP for DR energy management in the wireless part. For the web services, the browser/server (B/S) software architecture has been implemented to replace the previous client/server (C/S) architecture in our work, which was able to meet the current global network of open and interconnected devices. Moreover, the production planner or facility manager can access the

EMS, EMA and MCS by using these web services.

This proposed system can accommodate many types of DERs including solar panels, wind turbines and wasted heat recovery generated in the facility. Here, we only include a simulation model of a solar EGS in the experimental scenario because the size of the experimental model is small and the power consumption is limited. The structure of the system was composed of the simulation component and the physical component.

1. The Software Simulation

We used a software simulation to realize the PowerStation, the EGS and the ESS in energy management system.

(1) PowerStation

The PowerStation was realized by using a high-performance personal computer, on which a virtual utility supplier was built. It periodically provided day-ahead dynamic electricity prices to the EMS.

(2) ESS

A virtual ESS was developed as an in-facility DER using a high-performance personal computer (PC), which was developed using a software model. Here, the energy manager part of the ESS performs an internal energy management function to control the electrical storage device, which follows the EM model defined in the FSGIM.

(3) EGS

A virtual in-facility solar EGS was also developed using a software model on the PC. According to the FSGIM, energy is generated by the EGS Generator model, whereas the energy management function is performed by the EGS EM model for the solar EGS.

2. The Physical Component

This section describes the physical component of the developed platform.

(1) EMS

In this study, the EMS was developed using a PC. Based on day-ahead electricity prices, the EMS runs the DR algorithm to determine the optimal operating points of the STs and the operating status of the DERs to shift the electricity demand from peak to off-peak demand periods, and transmits the control information to the EMA and MCS.

(2) EMA

In this study, the EMA was developed using a PC. The EMA is acted as the energy manager agent for monitoring the electricity consumption and controls the electric load of each task.

(3) MCS

In this study, the MCS was built using a PC, which communicates with the EMA and EMS using Ethernet as a backbone network. A 6LowPAN border router was also connected with the MCS using the Serial Line Internet Protocol (SLIP) for enabling communication between the backbone network and the wireless field network. In the wired part, the communication is based on HTTP over TCP, whereas in the wireless part, communication is based on CoAP over UDP. Therefore, a mapping function for wireless and wired protocols was implemented in the MCS.

(4) Wireless field networks

In our experimental facility, we used 6LoWPAN as the wireless communication network. In each field device, we used DC motors as industrial loads, because motors are one of the most commonly used pieces of industrial equipment. Fig. 10.6 shows that the industrial process side included two tasks: NST and ST. The NST consisted of two motors, which were regarded as NSEs, and the ST consisted of three motors, which were regarded as representing each of the three categories (NSE, SE, and CE).

3. The Web-service Implementation with XML Based on FSGIM

The Web service is a mechanism for exchanging data between different systems developed by different parties. In this study, the Web-service implementation made the exchange data in a system-independent way. Here, FSGIM was implemented to standardize the data formats. XML was used to specify the FSGIM as a metadata exchange technology. HTTP/CoAP used XML as their data format. As a general purpose document format, XML provides a structured mechanism to encode machine-readable information.

The structure and semantics of the FSGIM are defined using the Unified Modeling Language (UML). UML diagrammatic conventions can define the normative relationships and multiplicity details of the FSGIM.

Additional attributes and relationships were added for the EMS, EMA and MCS in the industrial facilities. Then we made the updated FSGIM UML file using the Enterprise Architect (EA) tool. For each component, the EA tool could convert the UML class to XML schema. An XML schema is a description of a type of XML document, typically expressed in terms of constraints on the structure and content of documents of that type, above and beyond the basic syntactical constraints imposed by XML. XML is an instance of an XML schema. Each device may have more than one XML file.

On the wireless side, the limited resources in field nodes regarding processing capacity, power supply, and communication bandwidth necessitated the use of lightweight mechanisms. Representational state transfer (REST) is a lightweight instantiation of the Web services concept, which is particularly well suited to the properties of smart objects. In this system, the RESTful architecture was implemented. With this instantiation of REST, CoAP requests were used to transfer representations of resources between clients and servers.

Table 10.1 Parameters of devices for non-schedulable task

Operating Point	Non-shiftable Device (motor 1) r/min	Non-shiftable Device (motor 2) r/min	Power Consumption (mWh)	Consumption Rate of B(unit)	Consumption Rate of C(unit)	Production Rate of D (unit)
1	1200	600	620	60	120	100

Table 10.2 Parameters of devices for schedulable task at each operating point

Operating Point	Non-shiftable Device (motor 3) r/min	Controllable Device (motor 4) r/min	Shiftable Device (motor 5) r/min	Power Consumption (mWh)	Consumption Rate of A (unit)	Production Rate of B (unit)
1	600	0	0	210	30	20
2	600	600	0	420	60	40
3	600	1200	0	635	90	60
4	600	1800	0	940	120	80
5	600	1800	1200	1365	150	120

Table 10.3 Parameters of Intermediate product B and ESS

Parameter	Description	Value
B_Cap	Intermediate product storage capacity	500 units
B_Init	Intermediate product initial storage	50 units
I_{ess}	ESS initial storage	0 mWh
C_{ess}	ESS storage capacity	2500 mWh
θ_c	ESS maximum charging rate	600 mW
θ_d	ESS maximum discharging rate	600 mW
γ_c	ESS charging efficiency	0.9
γ_d	ESS discharging efficiency	0.9

V. Experimental Scenarios and Results

1. Experimental scenarios

As shown in Figure 10.6, we developed a STN model of the energy-management platform, which included four state nodes (input A, input C, intermediate product B and final product D) and two task nodes (schedulable task and non-schedulable task). In this model, the ST consumes input A to produce intermediate product B, and the NST consumes intermediate product B and input C to produce final product D.

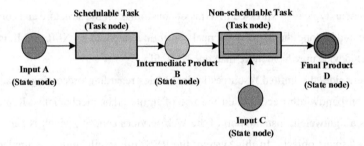

Figure 10.6 The STN model of the experimental scenarios

Tables 10.1 and 10.2 list the parameters of the NSTs and STs, respectively. The NSTs have only one operating point, specified by parameters such as the speed of each motor, power consumption, and the consumption and production rate of the related state nodes. The STs have five operating points, with the same parameters defined for each operating point.

In the experimental scenarios, the storage for input A, intermediate product B, and input C were modeled using software. The storage capacity and initial storage of intermediate product B were set to 500 and 50, respectively (Table 10.3).

In the experimental scenario, IESs were considered by using an ESS and a solar EGS, both present inside the facility. For the ESS parameters, initial storage was 0 mWh, storage capacity was assumed to be 2500 mWh, maximum charging and discharging rates were set to 600 mW, and the charging and discharging efficiency coefficients were set to 0.9 (Table 10.3). It was assumed that the rate of electricity charged/discharged during a given time interval could be continuously controlled between zero and the maximum charging/discharging rate.

There can be many different forms of facility energy generation system, such as solar, wind turbine and wasted heat recovery. In this experimental scenario, a solar EGS is used for software simulation. Figure 10.7

shows the predicted energy generated by the solar EGS during each time interval. The energy generation rate was high during the peak sun hours, while no energy was generated through the night.

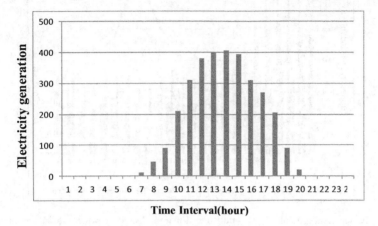

Figure 10.7   The predicted quantity of electricity generated by the solar EGS during each time interval

2. Evaluation of the Implementation

A wireless field network was implemented using a CC2538 microprocessor with the Contiki operating system. Contiki provides effective low-power communication ability and supports fully standard IPv6 and IPv4, as well as 6LoWPAN. The Contiki operating system has also been demonstrated as an easy-to-use platform for building CoAP-enabled DR energy management.

For the meter, a CoAP GET request, which is issued by the CoAP client, retrieves energy consumption from the meter. The CoAP server, which runs on the meter, sends the CoAP response with the energy consumption using XML structure to the CoAP client. For SE and CE of the load instances, a CoAP PUT request, issued by the CoAP client, sets the curtailment rating level attribute of the load, based on which the CoAP server controls the operating status of the load. The *Curtailment Ratings Level* attribute can also receive a CoAP GET request from the CoAP client with XML structure.

According to the dynamic change of price in each time interval, NST operated at the same operating point, while ST operated at different operating points. When the price was low, ST was scheduled to operate at a high energy consumption operating point, thereby increasing demand. When the price was high, ST was scheduled to operate at a low energy consumption operating point, thereby decreasing demand.

Figure 10.8 shows the energy demand under three different scenarios. In the fixed-price scenario, the electricity price was constant over all time intervals and was set to be equal to the average of the dynamic price. Compared with the fixed-price scenario, the scenario with DR consumed more electrical energy during low-price periods and less electrical energy during high-price periods, and thus the total cost was reduced. In the case of the scenario with DR and ESS, the gap in energy demand between its maximum and minimum becomes larger. Among the three scenarios, because ESS stored energy when prices were low and supplied energy when prices were high, the scenario with DR and ESS consumed the most electrical energy during low-price periods and the least electrical energy during high-price periods. With DR and ESS, the total cost was reduced even further, not only by shifting demand but also by managing IESs. The total electrical energy costs with DR and ESS decreased by 19.64% in comparison to the fixed-price scenario.

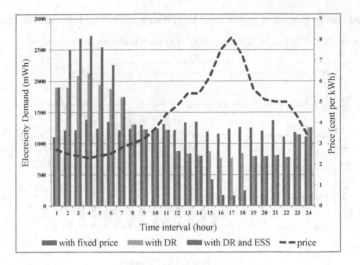

Figure 10.8　Experimental results energy consumption and electricity price with DR and ESS included

Figure 10.9 shows the energy demand under three additional scenarios. In the case of the scenario including DR and EGS, the electricity demand decreased further because the solar EGS generated electricity during time intervals 7 to 19. The EGS reduced energy costs by generating additional electricity using solar power. The total electrical energy costs when DR and EGS were included decreased by 24.22% in comparison to the fixed-price scenario.

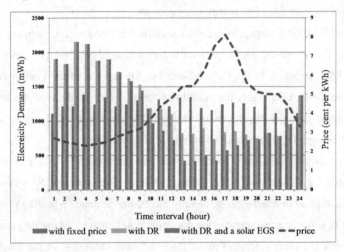

Figure 10.9　Experimental results of energy consumption and electricity price with DR and EGS included

Figure 10.10 shows the energy demand under a further four different scenarios. In the case of the scenario with DR, ESS and EGS, the ESS reduced energy costs because it stored electricity when prices were low for use when prices were high, and the EGS reduced energy costs by generating additional electricity using solar power. The electricity demand was negative during time intervals 16 to 17, which indicates that the facilities may sell surplus electricity to the grid to make a profit. The total electrical energy costs when DR, ESS and EGS were included decreased by 30.48% in comparison to the fixed-price scenario.

Table 10.4 shows the total energy cost for different cases (fixed price corresponds to the electricity price in each time interval being fixed, and equal to the average of the dynamic price).

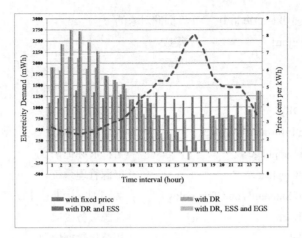

Figure 10.10 Experimental results of energy consumption and electricity price with DR, ESS and EGS included

Table 10.4 Total energy costs of experimental scenarios

	Fixed price	DR price	DR price with ESS	DR price with EGS	DR price with ESS and EGS
Costs (cent $)	1293.01	1136.69	1039.03	979.76	898.94

VI. Conclusion and Future Work

This paper focuses on proposing an IoT-based communication framework with a standard common information model. The application of the FSGIM to IES in industrial facilities is discussed. Current results show that the IoT-based architecture and FSGIM based information models can play an important role in providing interconnectivity and interoperability between devices and equipment, which guarantees simpler implementation, installation, operation and management of integrated energy management systems in a facility. It also offers a simple and common strategy for exchanging energy-related information between the entities in a facility.

In this study, we developed an IoT-based energy-management platform with common information models for industrial facilities. The integrated energy system architecture and related cases for utilizing different types of energy (EGS, ESS and utility side) are discussed. The IoT-based energy management system includes EMS, EMA, MCS, an industrial Ethernet backbone network and a wireless field network.

The results demonstrate that energy demand could be shifted from peak to off-peak demand periods, leading to significant reductions in overall cost. Additionally, the IoT-based technologies allowed the system to be implemented with less time, effort, complexity, and cost.

Future work will extend and implement the DR scheme to support the real-time energy price in the Cyber Physical System (CPS)-IoT platform. The IoT-based energy-management platform will be extended by developing the actual implementation of IESs, such as real rechargeable batteries, solar panels, wind turbines and wasted heat recovery. We also plan to extend the application to real industrial processes.

VII. New Words

interconnectivity ['ɪntəkənek'təviti]   n. 互连性，互联性，互联互通

interoperability ['ɪntərɒpərə'bɪlətɪ]   n. 互用性，互操作性，互通性

object-oriented ['ɒbdʒikt-'ɔːrɪentɪd]   *adj.* 面向对象的

integration [ˌinti'greiʃən]   *n.* 集成

load [ləud]   *n.* 负载

Ⅷ. Phrases

demand response (DR)	需求响应
Facility Smart Grid Information Model	设施智能电网信息模型
Extensible Markup Language	可扩展标记语言
energy management system	能源管理系统
National Institute of Standards and Technology	美国国家标准技术研究所
energy service interface	能源服务接口
canonical data model	规范化数据模型
common information model	通用信息模型
state-task network	状态任务网络
distributed energy resources	分布式能源
Industrial Ethernet	工业以太网
Utility Power Station	公用供电配置
Energy Manager System	能源管理系统
Energy Manager Agent	能源管理代理
Monitoring and Control System	监视与控制系统
Non-shiftable Equipment	不可调的设备
Controllable Equipment	可控的设备
Shiftable Equipment	可调的设备
Energy Storage System	能源存储系统
Electricity Generation System	发电系统
Use cases	用例
mixed integer linear programming	混合整数线性规划

Ⅸ. Abbreviations

LLNs   low-power and lossy networks

TCP   Transfer Control Protocol

UDP   User Datagram Protocol

HTTP   Hypertext Transfer Protocol

SNMP   Simple Network Management Protocol

CoAP   Constrained Application Protocol

UpnP   Universal Plug and Play

DHCP   Dynamic Host Configuration Protocol

SLP   Service Location Protocol

EMIX   Energy Market Information Exchange

Passage C Smart Grid Applications

The American Council for an Energy-Efficient Economy reviewed more than 36 different residential smart metering and feedback programs internationally. This is the most extensive study of its kind (as of January 2011). Their conclusion was: "To realize potential feedback-induced savings, advanced meters [smart meters] must be used in conjunction with in-home (or on-line) displays and well-designed programs that successfully inform, engage, empower and motivate people." There are near universal calls from both the energy industry and consumer groups for a national social marketing campaign to help raise awareness of smart metering and give customers the information and support they need to become more energy efficient, and what changes they must make to realize the potential of proposed smart meters.

Ⅰ. Australia

In 2004, the Essential Services Commission of Victoria, Australia (ESC) released its changes to the Electricity Customer Metering Code and Procedure to implement its decision to mandate interval meters for 2.6 million Victorian electricity customers.

The rollout commenced in mid-2009 and was completed at the end of 2013. Meters installed in Victoria have been deployed with limited smart functionality that is being increased over time. 30-minute interval data is available, remote cut-off and start-up energization is available, and the Home Area Network will be available for households. In May 2010 it was reported that the program was expected to cost $500 million more than originally estimated by proponents, with a total cost of $1.6 billion.

The Victorian Government after initially halting the planned implementation of Time-of-Use tariffs for general consumers has now allowed their introduction from mid-2013. As shown in Table 10.5, Victorian metering charges increased by approximately $60 per meter per year after the introduction of AMI cost recovery from customers in 2010 and a projected increase to 125.73 by 2016–2017. By mid-July 2013, the first Smart Meter In-Home Displays were being made available to Victorian consumers. At the beginning of 2014 there were three approved Smart Meter In-Home Displays directly available to consumers.

Table 10.5 Annual meter charge increases with smart meter costs in 2010 and projections to 2017 ($)

Distributor	2005	2006	2007	2008	2009	2010	2011	2012	2013	2015	2016 2017
SP AusNet	17.49	17.49	17.49	17.49	17.49	86.1	93.83	101.02	108.75	117.08	126.04
United Energy Distribution	6.60	6.60	6.60	6.60	6.60	69.21	89.18	99.57	107.62	116.33	125.73
Jemena Electricity Networks	12.87	12.87	12.87	12.87	12.87	134.63	136.7	155.84	159.86	162.34	164.88
Citipower	15.20	15.20	15.20	15.20	15.20	104.79	108.4	93.38	95.26	97.17	99.13
Powercor	17.20	17.20	17.20	17.20	17.20	96.67	105.35	92.72	93.91	95.12	96.34

II. Using Smart Grid Technologies to Modernize Distribution Infrastructure in New York

Under the American Recovery and Reinvestment Act of 2009, the U.S. Department of Energy and the electricity industry have jointly invested over $7.9 billion in 99 cost-shared Smart Grid Investment Grant projects to modernize the electric grid, strengthen cyber security, improve interoperability, and collect an unprecedented level of data on smart grid and customer operations.

Con Edison serves about 3.3 million residential, commercial, and industrial customers in New York City and surrounding areas, operates more than 700 MW of electric generation, and manages an electric distribution system that is 86% underground. Con Edison's electric service territory includes a mix of high-density urban loads, such as those found in mid-town Manhattan, and moderate-density urban and suburban loads found in other boroughs and surrounding counties.

Con Edison's SGIG project installed distribution automation technologies and systems to improve electric reliability, remote monitoring, operator decision support, asset utilization and capacity management, energy savings and efficiency, reactive power management, and power quality and substation battery monitoring.

The total budget for the entire project is $272.3 million, including $136.2 million in SGIG funding from the U.S. Department of Energy (DOE) under the American Recovery and Reinvestment Act of 2009. The subproject in this case study involves 4kV grid modernization and includes installation of 449 pole mounted distribution capacitors; 111 digital LTCs; power quality and battery monitoring technologies; and development of a 4kV grid model for enhanced load flow analysis. The budget for this subproject is almost $20 million or about 7% to the project's total budget.

The portion of Con Edison's distribution system that uses 13kV and 27kV equipment has a greater percentage of substations and feeders currently equipped with automated digital controls and access to SCADA systems. In contrast, the 4kV grids use analog systems with electro-mechanical controls and typically serve areas where load growth has been modest. Replacing the 4kV grids with higher voltage systems is cost prohibitive in New York due in part to the high cost of real estate. Prevalent throughout the U.S. and other parts of the world, Con Edison's 4kV grids represent about 34% of distribution circuits, 16% of customers, and 11% of system peak demand.

Reduction in electrical line losses and reactive power management equates to less electrical generation required to supply system demand. System losses and poor reactive power management mean that power plants need to generate greater amounts of reactive power, which can cause higher fuel consumption and air emissions. Power factors on the 4kV grid can vary throughout the day as inductive loads, such as air conditioning, rise and fall. Deployment of automated capacitors provides local sources of reactive power and reduces the amount that would be needed overall.

Con Edison's 4kV grids are supplied by several unit substations that are typically equipped with a single power transformer that, in some locations, experience circulating reactive power flows caused by voltage imbalances. These circulating flows are harmful to transformers, cause high internal heating, and result in de-ratings. To reduce these potentially damaging effects Con Edison has installed new digital load tap changer (LTC) controls at 111 of the 4kV unit substations and implemented improved control methods. Distribution system operators can control each substation transformer LTC controllers through their SCADA system.

Ⅲ. **New Words**

residential [ˌezɪ'denʃ(ə)l]	*adj.* 住宅的；与居住有关的
feedback ['fiːdbæk]	*n.* 反馈；成果，资料；回复
conclusion [kən'kluːʒ(ə)n]	*n.* 结论；结局；推论
conjunction [kən'dʒʌŋ(k)ʃ(ə)n]	*n.* 结合；[语] 连接词；同时发生
commonwealth ['kɒmənwelθ]	*n.* 联邦；共和国；国民整体
budget ['bʌdʒɪt]	*n.* 预算，预算费　*vt.* 安排，预定；把…编入预算

Ⅳ. **Phrases**

American Council	美国理事会
smart meters	智能仪表
American Recovery and Reinvestment Act	美国复苏与再投资法案
Con Edison	联合爱迪生电力公司

Ⅴ. **Abbreviations**

ESC	Essential Services Commission	基本服务委员会
SGIG	Smart Grid Investment Grant	智能电网投资补贴
DOE	Department of Energy	能源部

Exercises

Translate the following sentences into Chinese or English

1. Smart grid is one of the most potential field for big data application, including data resources and characteristics, application value presentation of big data in smart grid, difference in research approach between big data and traditional research method.

2. On the basis of domestic and international smart grid theoretical research and construction practices, starting from the analysis of basic power grid mode, combining with the external drive for smart grid from economic and social aspects, a development mechanism analysis methodology is put forward which considers the factors both inside and outside of power system.

3. 目前，智能电网已成为世界电网发展的大趋势，符合社会和经济发展的必然要求。

4. 家庭能源管理系统是智能电网在居民侧的延伸，是智能电网领域的研究热点之一。

参考译文 Passage A 智能电网介绍

Ⅰ. 如何定义智能电网

智能电网没有统一的定义。但智能电网承载了一系列的期望，必须满足广泛的新要求。智能电网必须优化当前的电网网络，它配有先进的传感执行器和高度安全的网络基础设施，以提高电网的效率、性能和可靠性，以及支持广泛的新服务。例如，对耗电量描述更好的认知，插电式混合动力车的使用，分布式能源（如太阳电池板和住宅发电机设备以及智能家居的应用）。

智能电网是物联网的主要应用之一，互联网协议是其核心问题。在本章的几个例子中，大多数对智能电网的期望与要求都包含在智能网络对象之内：传感器（例如测量电流、电压、相位及无功功率）与执行器（例如断路器等）用于有效地监测和控制电网，智能仪表用于测量功率消耗，大量的智能设备都可以通过专门的能源管理设备与家居、建筑及工厂进行通信，这样使得电网具有高效能源管理的能力。

图10.1描述了一个典型的电网结构，从发电到家庭、建筑：电是由工厂产生的，然后通过配电网分配给终端用户。格状网的特点在于层级结构。高压线与（主）变压站相连接，其电压通过杆顶（美国）或二次变电站（欧洲）在降到低压前，会被降低到一个平均值。最后，电力被传送给终端用户，用户使用智能仪表监控能源（并支持其他功能）。

Ⅱ. 智能电网的特征

智能电网代表了当前的和建议的应对电力供应挑战的全套解决办法。由于各种各样的因素，虽然有大量相互竞争的分类法，但对于普遍的定义没有达成一致。因此这里给出一种可能的分类。

1. 可靠性

智能电网将使用新技术，例如状态估计，在没有技术人员参与的情况下，可以提高故障检测，并提供网络的自我修复。这将确保更可靠的电力供应，并降低自然灾害或攻击的易损性。

虽然多路由被吹捧为智能电网的一个特征，但旧电网也有多路由的特征。输电网中的初始电源是利用放射模型构成的，之后的连接性通过多路由来保证，简称为网络结构。然而，这就产生了一个新的问题：如果当前电流或通过网络的相关作用，超过任何特定网络元素的限制，它可能会失败，并且当前流动会分流到其他网络元素，最终也可能失败，造成多米诺效应。对于电力中断，通过滚动停电或降低电压（降压）技术，可以防止这种用电限制。

2. 网络拓扑的灵活性

下一代传输和分配基础设施，能更好地处理可能出现的双向能量流动，允许分布式发电，例如，屋顶的太阳能系统以及燃料电池，可以从电动汽车、风力涡轮机、泵水力发电以及其他来源提供或获取电力。

传统电网被设计为单向流电力，但如果一个本地子网产生力多于它的消耗，反向流就可以提高安全性和可靠性。智能电网的目标就是管理这些情况。

3. 效率

智能电网技术的部署有利于全面提高基础设施效率，特别是需求方面的管理，例如在短期电价峰值关闭空调、通过无功优化降低配电线路电压、消除重复读取表的数据、通过采用先进计量基础设施系统的数据加强停电管理。取得的效果如下：减少了传输和配电线路的冗余，更高效地使用了发电机，降低了电力价格。

4. 可持续性

智能电网灵活性的改进，允许高度可变的再生能源有更大的渗透力，如太阳能发电和风力发电，即使没有能量储存的增加也能实现。目前的网络基础设施建设不允许多个分布式的馈入点，通常情况下，即使本地（分布）允许一些馈入，传输级基础设施也不能适应它。如由于多云或突发的天气状况，分布式发电会出现快速波动，电力工程师面临着重大挑战，需要通过调变多个可控发电机的输出，例如，燃气轮机和水轮发电机，确保稳定的输出功率水平。因此智能电网技术，是大量可再生电力的必要条件。

Ⅲ. 为什么智能电网如此重要

电力被引入千家万户，全球电网正发生巨大变革。现在的电力基础设施，正在迅速地发展，以克服其局限性，虽然没有很强的现代化，但它已经很好地为我们服务。能源生产正从集中式向分布式转变，形成双向功率流，从放射性结构转变为网状拓扑结构。澳大利亚资源，能源和交通运输部，最终得出结论，现有的电网基础设施需要升级和重新设计，以方便集中式向分布式发电的转变。

当前的电网必须扩大，支持分布式发电的不断增加，并且着眼于集成分布式发电的本地化和全系统的影响，包括高比例的可再生能源。由于天气的变化性和有限的可预测性，可再生能源可以产生电压波动和配电线路中的谐波。因此，一个健壮的、高速的、双向的通信网络为智能电网实施奠定了很好的基础。该网络将覆盖当前电网，并提供前所未有的控制和监测功能，同时还具有自治功能和智能算法。

电网运营商不愿意建立整个电信网络，用来控制和监测智能电网，而倾向于使用当前技术（如LTE）的私有无线网络。然而，在通信业务量已经很高的情况下，这使通信业务量变得更高。研究表明，通过优化通信资源，增加智能电网数据分类的影响会被最小化，并保留其QoS性能。

Ⅳ. 什么是智能抄表

智能抄表将为消费者与公用事业之间提供通信手段。这将使其他技术一体化，如需求响应。实时消费水平的数据可转移到公用事业，并会使消费者监测他们的电力消耗，并采取措施以减少使用。此外，智能抄表可提供电力的实时电价或间接负荷控制，这被称为动态价格反应，包括：

1. 分时电价（ToU）收费表

这项计划鼓励消费者将消费从高峰时间转移到非高峰时间。这是一个非动态的收费表，实际上这样一个收费表的大范围整合可能反过来改变高峰时间。

2. 实时定价（RTP）

市场的电价每小时都在变化，或者在有些市场半个小时变化一次。实施定价的程序员提供一种每小时变化的收费表，以反映市场供需和电价的变化。激励消费者在电价高时少用电，电价低时多用电。

3. 谷峰分时电价（COP）或者称为尖峰电价（CPP）

实时传输协议（RTP）对于居民用户来说是不可行的，而尖峰电价（CPP）是一个合理的选择。尖峰电价（CPP）收费表在系统压力期间以可分派的高价或"尖峰"电价增大时间不变的费率结构或分时电价结构。每年数量有限的自由支配日，或者系统或市场条件满足预设条件时，尖峰电价可能会出现。参与客户一般提前一天收到分派的高价通知，在某些情况下，提供自动控制技术以支持高效的负荷下降。因为尖峰电价（CPP）费率的价格都是预设的，所以尖峰电价（CPP）没有实时传输协议（RTP）经济效益好；但是，因为它减少与实时传输协议（RTP）相关的潜在地巨大价格风险，这一相同的特点也使尖峰电价（CPP）更具吸引力。

智能抄表的另一个主要好处是，可用于一些公用事业的偷电检测。由于通信是实时发生的，任何测量设备的干扰，或绕过该表可发送到电厂。

V. 智能抄表目前的通信标准

关于智能抄表通信标准的工作已经进行了很多年，但由于最近几个通信行业的技术进步，可用的数据传输带宽已成倍增长。

智能抄表的现行通信标准，可分为有线和无线两种主要的通信方式。然而，它们是由一些系统相互交织在一起的，这些系统采用混合的技术以融合低功耗的射频通信与电力线载波。

1. 有线通信

电力线载波通信（PLC）：有线智能抄表通信中最主要的技术是电力线载波通信，也称为电力线载波。自从1980年，电力线载波通信就已经发展了，它能与已安装仪表的直接连接，是与电表通信的首先技术。然而，由于通信信号直接施加在输电线上，导致信号辐射到电线周围环境的技术问题。高带宽的电力线载波通信（PLC）已被证明会干扰用于地理定位的DGPS带。

由于缺乏不同的欧姆可用分离电路，带宽要求也强调通信介质的不足。大多数传输线是三相的，且通信仅限于低频带，两点之间频道的潜在总带宽是有限的。

研究电力线载波通信（PLC）上时变信道的影响，要在数十兆赫的频率响应下功夫。但是，这些频率的衰减对长距离的传输线造成信道串扰的问题。为了保证通信信道的鲁棒性，必须采用信道编码单元。使用线性分组码选择适合的频率，使两节点编码相通，正交频分多路复用技术（OFDM）可以降低信噪比。

电力线载波通信信道的进一步优化，可以来自电力分配系统中互联节点的多径行为的深入分析。根据从低电压变电站到用户前端的信号传播的衰减效应，可以看出包括延迟效应，在本质上是循环的，这导致任何基于相位的编码方案效率低，尤其是由于非线性延迟函数。

2. 无线通信

无线通信将其自身表示为一种使用有线技术的替代选择，实施无线的优势主要是降低基础设施的成本，因为分离节点之间的无线系统要求没有物理布线。然而，当数据通过物理空间传播，任何人在发送节点的范围内可截取数据，这是令人担心的事情。

（1）低功耗无线：应用测量技术时，显然需要低功率收发器。如果需要采集在传输网络中实时监测的节点消耗或生产电能，那么，可能需要监测连接到家庭配电网络的每个设备，以提高测量、监测和控制能力。

（2）ZigBee（IEEE 802.15.4）：虽然ZigBee在噪声环境下的鲁棒性越来越受关注，但ZigBee仍是低功率运行选择的通信标准。ZigBee节点在802.11/b/g干扰下的通信信道可能被毁灭，让它完全无法使用和误判为不切实际的"永远"连接。

（3）Z-Wave：一种源代码不开放的，与ZigBee类似的技术，声称由于其操作在800MHz范围内远低于802.11/b/g的频带有更好的抗干扰鲁棒性。

（4）网状网络：网状网络协议消除系统中每个节点需要在主节点的传输范围内的需要。数据包可与相邻的节点交换，随后可以把信息传给他们的邻居，到数据包到达其预定的收件人。

节点间数据包的传递允许数据源和目的地之间有远距离，然而，在有许多连接节点的网络中，一个数据包在邻居之间传递会来回循环并被转移回节点周围的发端。这不可避免地会导致通信信道中的开销，可以减少可用数据负载的有效带宽。

VI. 智能抄表的应用

当前，一些欧洲国家智能电表已经代替了旧的电能表。在意大利自治市（罗马、米兰、都灵、布雷西亚、帕尔马、维罗纳、的里雅斯特、博洛尼亚等）拥有大经销商（ENEL）和中型经销商，还有更小的当地经销商出售智能电表。对于ENEL，智能电表是强制性的，而其他经销商则继续使用旧的电表系统。通常，电表是属于经销商的，用户不允许私自买电表。ENEL从2002年开始，5年内设法安装了近2700万个电表，投资了22亿欧元。不同类型的用户都安装了智能电表，截至2006年12月，ENEL的电表超过95%都是智能电表。

EDF是欧联最大的电力经销商，作为EDF的子公司，ERDF在法国拥有3300万用户。在2008年7月，EDF宣布在全国范围内推出3500万智能抄表的第一个阶段。这个由国际信息技术公司源讯集团领导的团体将实施一个试点工程，这个工程在法国包含30万电力线载波通信仪表和7000个数据集中器，并预期在2010年初首次展出此项目。

20世纪90年代，SCE在自动抄表方面进行了几次试点。但这些试点在当时没有实现大规模的实施。然而，因其为后来试点以及最终实施智能抄表系统识别了重要的需求，所以它是很有价值的。在21世纪，加州公共事业委员会（CPUC）呼吁公共事业部门调查智能电表技术的使用。这个要求使SCE开始了智能连接方案。此方案计划为所有的用户安装智能电表，而且将它们与公共设施的用户保护和激励计划集成在一起。

参考答案

1. 智能电网是大数据最重要的应用领域之一，包括：数据源及数据特征，大数据应用于智能电网的价值体现，大数据与传统研究方法的不同之处。

2. 在国内外智能电网理论研究与建设实践基础上，从分析电力系统的基本模式出发，结合经济社会对于智能电网的外部驱动作用，提出了一种内外结合的智能电网发展机理分析方法。

3. At present, smart grid has become a general development trend of the grid, it can meet the development requirements of the society.

4. Home energy management system (HEMS) is an extension of smart grid in residential sector, and it is a hot topic of smart grid.

物联网国际标准组织

从电子标签（Radio Frequency Identification，RFID）、机器类通信（Machine to Machine，M2M）、传感网（Sensor Network，SN）、物联网（Internet of Things，IoT）到泛在网（Ubiquitous Networking，UN），国外标准组织开展了大量的物联网相关标准工作。

国际电信联盟远程通信标准化组织（ITU-T for ITU Telecommunication Standardization Sector，ITU-T）创建于1993年，其中文名称是国际电信联盟通信标准化组织，是国际电信联盟（ITU）旗下专门制定远程通信相关国际标准的组织。早在2005年，ITU-T就开始进行物联网研究。2011年5月ITU-T召开了第1次物联网全球标准化倡议活动。自此ITU-T正式开始了一系列物联网标准的制定工作。到目前为止，ITU-T已经发布了物联网系列标准，如Y.2060。

3GPP（3rd Generation Partnership Project，第三代合作伙伴计划）作为移动网络技术主要的标准组织之一，其关注的重点在于增强移动网络能力以满足物联网应用所提出的新需求，是在网络层面开展物联网研究的主要标准组织。目前，3GPP针对M2M的需求主要研究M2M应用对网络的影响，包括网络优化技术等。具体研究范围为：只讨论移动网内的M2M通信，不具体定义特殊的M2M应用。

因特网工程任务组（Internet Engineering Task Force，IETF），成立于1985年底，是全球互联网最具权威的技术标准化组织，主要负责互联网相关技术规范的研发和制定，当前绝大多数国际互联网技术标准出自IETF。IETF中的多个工作组，如CoRE工作组、6LoWPAN工作组等，涉及物联网的应用层（ISO-OSI模型）和网络层（ISO-OSI模型）标准。

IEEE（Institute of Electrical and Electronics Engineers，电气和电子工程师协会）自成立以来一直致力于推动电工技术在理论方面的发展和应用方面的进步，现如今也开始着眼于物联网标准制定工作，期望在物联网领域取得一定优势，IEEE先后成立了IEEE 2413（物联网体系架构）、IEEE 1451（智能接口）与IEEE 802.15等工作组来从事物联网的相关工作。IEEE 2413主要针对物联网体系架构进行研究，于2014年底成立。IEEE 1451主要研究工作集中在传感器接口标准方面，发布了IEEE 1451.1-IEEE 1451.5系列标准协议。IEEE 802.15主要规范近距离无线通信，于2003年10月1日如发布了第一版本标准，即为IEEE 802.15.4-2003，随后又陆续发布IEEE 802.15.4-2006，IEEE 802.15.4-2011对先前版本进行完善与改进。

ZigBee联盟成立于2001年8月，是IEEE 802.15.4组织对应的产业联盟。ZigBee负责制定网络层到应用层的相关标准，针对不同的应用制定了相应的应用规范。其对应的物理层和链路层是在IEEE 802.15.4组织研究制定。Zigbee组织目前包含23个工作组和任务组，涵盖技术相关的工作组：架构评估、核心协议栈、IP协议栈、低功耗路由器、安全；以及应用相关的工作组：楼宇自动化、家庭自动化、医疗、电信服务、智能电力、远程控制、零售业务，还有与市场、认证相关的一些工作组。Zigbee目前发布了3个版本的协议栈规范，第一个Zigbee协议栈规范于2004年12月正式生效，于2005年9月公布并提供下载，称为Zigbee 1.0或Zigbee 2004；第二个Zigbee协议栈规范于2006年12月发布，此版本对Zigbee 1.0进行标准修订，为Zigbee 1.1版（又称为Zigbee2006）；第三个Zigbee协议栈规范于2007年10月完成，称为Zigbee Pro或Zigbee2007。

OMA（Open Mobile Alliance，开放移动联盟）始创于2002年6月，是由WAP论坛（WAPForum）

和开放式移动体系结构（Open MobileArchitecture）两个标准化组织合并而成。随后，区域互用性论坛（Location Interoperability Forum，LIF）、SyncML、MMS互用性研究组（MMS Interoperability Group，MMS-IOP）和无线协会（Wireless Village），这些致力于推进移动业务规范工作的组织又相继加入OMA。OMA终端管理协议即OMA DM协议是目前M2M移动终端管理的热门协议之一。目前已有OMA DM1.3和OMA DM 2.0两个版本。另外，为了支持资源受限设备的终端管理需求，OMA还制定了LightWeight M2M协议。

ISO/IEC JTC1下SWG5（Special Work Group 5，第五特别工作组）于2012年在ISO/IEC JTC1第二十七次全体会议上通过成立，SWG5的主要任务是致力于物联网体系架构的研究。此组织由四个特设小组（AHG）组成，每个小组的任务各司其职，比如AHG1的主要任务就是去制定一个大众化的物联网概念，AHG2的任务就是去分析物联网对市场的要求。2014年，ISO IEC JTC1 WG10（Working Group 10）物联网工作组在ISO/IEC JTC1全体会议上通过成立，其主要目标是着手于物联网基本标准的制定，以便为物联网其他标准的发展奠定一个夯实的基础。制定物联网词汇的形式和定义、制定物联网的参考架构和基础协议等等都是WG10物联网工作组的任务。ISO/IEC JTC1/WG 7（传感网工作组），由中、美、德、韩四国推动并成立，其主要任务是开展传感网领域标准的制定。国际主要物联网相关标准组织如图10.11所示。

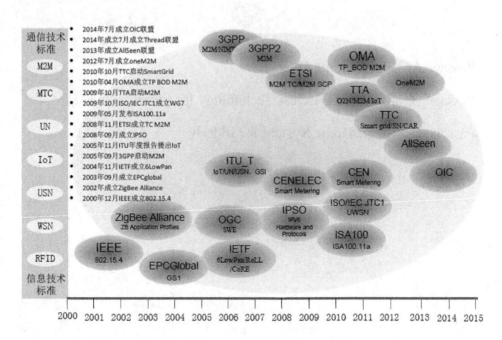

图10.11　国际主要物联网相关标准组织

Unit 11

Smart Home

Passage A   Smart Home Overview

Passage B   EnOcean Wireless Technology for Smart Home

Passage C   Smart Home System for Elderly

Passage A Smart Home Overview

I. The introduction of the Smart Home

A smart home can be viewed as an environment in which computing and communications technologies employ artificial intelligence techniques to reason about and control our physical home setting. In a smart home, sensor events are generated while residents perform their daily routines. A smart home is an ideal environment for performing automated health monitoring and assessment. Using this setting, no constraints are made on the resident's lifestyle.

Smart home promises the potentials for the user to measure home conditions (e.g., humidity, temperature, luminosity, etc.), manipulate home HVAC (heating, ventilation and air conditioning) appliances and control their status with minimum user's intervention.

A smart home system consists of subsystems and devices like home automation, CCTV cameras, fire alarms, audio streaming and infotainment, energy management appliances and so forth. Figure 11.1 shows the components of smart home systems.

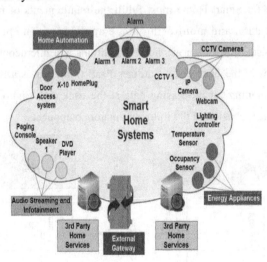

Figure 11.1 Components of smart home systems

II. The history of the Smart Home

A remote control for moving vessels and vehicles was first patented by Nikola Tesla in 1898. While many home appliance have existed for centuries, the self-contained electric or gas powered appliances became viable in the 1900s with the introduction of electric power distribution. In the early 1900s, electric and gas appliances included washing machines, water heaters, refrigerators and sewing machines. In the Post-World War II economic expansion, the domestic use of dishwashers, and clothes dryers were part of a shift for convenience and increasing discretionary income.

The first micro processors emerged in the early 1970s and were used in embedded systems like calculators, and microcomputers. The Honeywell Kitchen Computer of 1969 was offered by Neiman Marcus for $10,000 ($63,730 in 2013), weighed over 100 pounds (over 45kg), and was advertised as useful for storing recipes. Reading or entering these recipes required the user to complete a two-week course just to learn how to program the device, using only toggle-switch input and binary light output. It had a built in cutting board and had a few recipes built in. No evidence has been found that any Honeywell Kitchen Computers were ever sold.

In 1975, X10 the first general purpose Home automation network technology was developed. It is a communication protocol for electronic devices. It primarily uses electric power transmission wiring for signaling and control, where the signals involve brief radio frequency bursts of digital data and remains the most widely available. Although higher bandwidth alternatives exist, X10 remains popular in the home environment with millions of units in use worldwide, and inexpensive availability of new components. By 1978, X10 products included a 16 channel command console, a lamp module, and an appliance module. Soon after came the wall switch module and the first X10 timer.

By 2012, in the United States, according to ABI Research, 1.5 million home automation systems were installed.

III. The system Architecture for Smart Home

A system architecture for Smart Home must fulfill the requirements of measuring home conditions, processing instrumented data, and monitoring home appliances. The proposed approach utilizes microcontroller-enabled sensors for measuring home conditions and microcontroller-enabled actuators for monitoring home appliances in the front end. It utilizes PaaS (Platform as a Service) and SaaS (Software as a Service) in Cloud computing for processing data at the back end. Figure 11.2 illustrates the system architecture for Smart Home. It consists of the following major components:

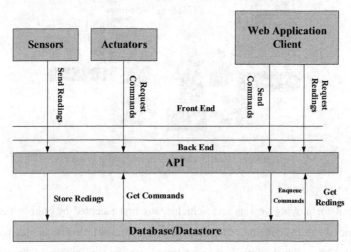

Figure 11.2 System architecture for Smart Home

- Microcontroller-enabled sensors: measure home conditions; the microcontroller interprets and processes the instrumented data.
- Microcontroller-enabled actuators: receives commands transferred by the microcontroller for performing certain actions. The commands are issued based on the interaction between the micro controller and Cloud services.

- Database/Data Store: stores data from microcontroller-enabled sensors and Cloud services for data analysis and visualization, and serves as command queue being sent to actuators as well.
- Server/API layer between the back end and the frontend: facilitates processing the data received from the sensors and storing the data in database. It also receives commands from the web application client to control the actuators and stores the commands in database. The actuators make requests to consume the commands in the database through the serer.
- Web application serving as Cloud services: enable to measure and visualize sensor data, and control devices using a mobile device (e.g., smart phone).

IV. Technical Overview of Smart Home Technologies

Various projects, laboratories and industrial showcases regarding the smart homes are all over the world; while some might be more advanced than others, they share many similarities. Depending on research groups, the objectives of smart home projects can vary from proving specific technological innovations, gathering usability information, validating testing results, to demonstrating the latest commercial technologies.

In terms of the quality of life, most of the available smart home technologies can be classified into the following three main clusters:

1. Safety Enhancing Systems
- Technologies that enable independent living, including fall detectors, personal emergency response systems (PERS) and medication management systems.
- Technologies that enable and facilitate the dispatch and delivery of services to individuals in their living setting.
- Technologies that make living environments safer.

2. Health and Wellness Monitoring
- Technologies to manage chronic diseases like cardiovascular disease, asthma, diabetes, COPD, and dementia.
- Technologies for medication management, secondary prevention, and "active" tele-health technologies that allow live interaction with the patient through conversations.
- Comprehensive and consumer-friendly Personal Health Records (PHRs) to manage care delivery across different sectors.
- Connecting and integrating different technologies through enhanced interconnectivity and the ability to exchange information between different information systems.
- The health and well-being technologies that foster a consumer-centered market.

3. Social Connectedness Systems
- Technologies helping seniors seek opportunities for social connectedness, expand their social network.
- Measuring social interactions, through sensor and other technologies, and providing feedback may give seniors an opportunity to identify social efficiencies and take more control of their lives.

V. New Words

appliances [əpˈlaɪənsɪz]   *n.* 器具，器械，装置

recipes ['resəpɪz]	n.	食谱；烹饪法
console [kən'səʊl]	n.	控制台，操纵台；演奏台；悬臂；肘托
visualization [ˌvɪʒʊəlaɪ'zeɪʃn]	n.	可视化；形象化，想像；目测
validating ['vælɪdeɪtɪŋ]	v.	确证；证实
clusters [k'lʌstəz]	n.	串；簇
interconnectivity [ɪntəkənek'təvɪtɪ]	n.	互联性；互连性；互联互通；连通性

VI. Phrases

data store	数据存储
safety enhancing systems	安全增强系统
fall detectors	跌倒探测器
medication management systems	药物管理系统
health and wellness monitoring	健康监控
secondary prevention	二级防御
social connectedness systems	社会联通系统

VII. Abbreviations

HVAC	heating, ventilation and air conditioning	加热、通风和空气调节
PERS	personal emergency response systems	个人应急反应系统
COPD	chronic obstructive pulmonary disease	慢性阻塞性肺病
PHRs	Personal Health Records	个人健康档案；人健康记录

Passage B　EnOcean Wireless Technology for Smart Home

Smart homes and home automation are ambiguous terms used in reference to a wide range of solutions for controlling, monitoring and automating functions in the home. Berg Insight's definition of a smart home system requires that it has a smartphone app or a web portal as a user interface. Devices that only can be controlled with switches, timers, sensors and remote controls are thus not included in the scope of this study. Smart home systems can be grouped into six primary categories: energy management and climate control systems; security and access control systems; lighting, window and appliance control systems; home appliances; audio-visual and entertainment systems; and healthcare and assisted living systems Icontrol company released a smart home report 2015. By and large, smart home is here – from connected toothbrushes to appliances and HVAC, consumers can buy connected products online, on shelves and from the service providers that bring cable, Internet and security systems to their homes. However, the recent excitement over the "Internet of Things" means consumer preferences have taken a backseat to the "cool factor" of new product launches.

EnOcean is the originator of patented energy harvesting wireless technology. EnOcean manufactures and markets energy harvesting wireless modules for use in building, smart home and industrial applications as well as for the Internet of Things.

The idea of EnOcean's innovative technology is based on a simple observation: where sensors capture measured values, the energy state constantly changes. When a switch is pressed, the temperature alters or the luminance level varies. All these operations generate enough energy to transmit wireless signals.

1. EnOcean Batteryless Wireless Technology

The batteryless wireless technology from EnOcean has a special position in the market. The components equipped with this technology not only work wireless but batteryless too. Due to Energy Harvesting, the wireless modules gain their energy from the environment, for example from motion, light or temperature differences. Instead of batteries, mini solar cells, electro-mechanic and thermal energy converters provide the energy needed. Through this self-sufficient operation, any maintenance usually needed with similar, battery-driven solutions, is no longer necessary. The radio transmission used has very low energy demand while simultaneously offering a large range. Outdoors, this range is about 300 meters, in buildings it lies at up to 30 meters. One reason for the low energy consumption is the short transmission period. Within just 1 millisecond (ms), the complete transmission- /reception process occurs. Not only does this save energy, the chance of collision between different telegrams is lowered. Two repeat-telegrams are sent within 30 ms in order to prevent erroneous transmissions.

Batteryless wireless technology combines the respective advantages of both wired and wireless systems: it works maintenance-free, guarantees reliable operation and is flexible and cheap to install.

EnOcean wireless is internationally standardized as ISO/IEC 14543-3-10. This standard is optimized for radio solutions with particularly low energy consumption and Energy Harvesting. Together with application profiles (EnOcean Equipment Profiles) from the EnOcean Alliance, the prerequisites for a completely interoperable and open radio technology, comparable with Bluetooth or Wi-Fi, are given. In this way, products from diverse manufacturers can work together in one system without a problem. In Europe, batteryless wireless technology uses the frequency channel 868 MHz, which is only approved for pulse signals. EnOcean exclusively offers its batteryless wireless technology to product manufacturers (OEMs), which then integrate it into their solutions. These partners are organized in the EnOcean Alliance, an independent initiative with over 350 members, who have already successfully brought more than 1,200 interoperable products into the market. The offer covers batteryless switches, intelligent window handles, temperature-, humidity-, or light sensors, presence detectors, as well as actuators, gateways, and switchboards, along with complete Smart Home systems. Since 2013, radiator valve attachments are also offered, which use the temperature differences present at radiators, in order to power communication and changes in stroke.

II. Guarantee of Flexibility and Interoperability

The term "Smart Home" does not stand for the obligatory networking of all eligible components. Moreover, it is a constantly expandable system, which the user can change according to individual needs and technical advancement. If at first a security system is installed, later on the system can be extended to include comfort functions, for example. Basic modules such as the control unit, but also single components, can be used continuously. Presence detectors from a security system can be also used with luminaries. Generally, all controls of a building technology, from the blinds to the heating control, are included. The prerequisite here is that all system components fall back on a standardized communication process. The international EnOcean wireless standard and the EnOcean Alliance's uniform application profiles make sure that all components from different integration manufacturers are compatible with each other and stay compatible in the future. Gateways guarantee the reliable exchange of information to other systems such as KNX, LON, ZigBee, GSM or Wi-Fi. This especially is a fundamental feature for the use of batteryless wireless technology in the smart home. Due to different requirements concerning data rates and energy consumption, many communication systems are used there. If sensors communicated via Wi-Fi for example, they would have to be supplied externally with electricity. On the other hand, the transfer of a video from a smartphone to a TV via the EnOcean radio protocol would take days. The core function of the batteryless wireless technology in a smart home is the connection of sensors and actuators until the very last meter. Other systems are responsible for the optimum transfer of information via other channels, such as from a gateway to a smartphone.

III. Networking Structures

A far-reaching networking of devices also brings the highest benefits of use. That's why all trades gravitate this way in their development. First of all, all components in one area are networked together. In the second step, a supervisory connection between the applications is created, for example between central control and smartphone. The figure 11.3 shows the step-by-step construction of this networking structure.

As you can see from the image, single components are networked differently, depending on which technology is best suited. But because a smart home binds all devices, different Bus-systems must also be connected with each other. If a user sits on his couch in the living room, he wants to be able to control the temperature via his smartphone, in addition to playing music on the HiFi and all that with one simple and comfortable application. Current products available on the market show that this technical problem can be realized without much effort. A good example would be the Vitocomfort 200 packet manufactured by TELEFUNKEN Smart Building and distributed by Viessmann. With it, the user can control devices from afar or read conditions such as the room temperature. The two standards, Wi-Fi and EnOcean-wireless are connected via gateway.

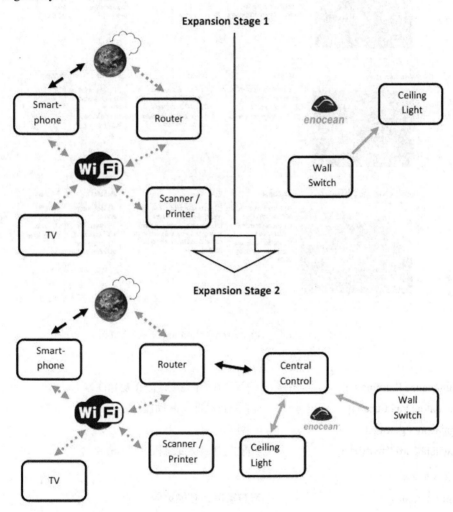

Figure 11.3 Change in the network structure

IV. Solutions

Figure 11.4 shows the solutions for smart home.

Ruby House – USA	Challenge	Solution	Benefits	Partners
	(1) Wireless lighting control (2) Wireless shutter control (3) Control via smartphones/PC	(1) Verve Living Systems light switches (2) Illumra hand-held switch (3) BSC remote control (4) Magnum energy metering (5) Gateway EnOcean to WiFi	(1) High flexibility (2) Complete interoperability (3) Individual programming (4) High comfort (5) High energy savings	(1) Magnum (2) Verve Living Systems (3) Illumra (4) BSC Computer
Weberhaus – Germany	Challenge	Solution	Benefits	Partners
	(1) A high-class, intelligent home control (2) Setting of new ecological standards (3) Integration of building automation and smart metering	(1) Four comfortable smart home models based on EnOcean solutions	(1) High flexibility (2) Easy installation (3) Cost-efficient (4) Suitable for radio-sensitive people	(1) Eltako (2) PEHA (3) Thermokon (4) BSC Computer (5) BootUp
Villa Torri – Italy	Challenge	Solution	Benefits	Partners
	(1) Central control of HVAC, shading and light (2) Remote access (3) Flexible system adaption	(1) Smart home system based on myGEKKO (2) WAGO actuators (3) EnOcean-based switches, occupancy sensors	(1) High comfort (2) High energy savings (3) Low installation costs	(1) myGEKKO (2) WAGO
DOMO LOGIS TEC – Belgium	Challenge	Solution	Benefits	Partners
	(1) Integration of KNX and EnOcean (2) System of cross-vendor solutions (3) Intelligent networking of several building areas	(1) Automated control of several energy efficiency measures (2) Network of geothermal energy, lighting, rain water and single room control depending on the actual situation	(1) High energy savings (2) High comfort (3) Easy control (4) Solution for individual needs	(1) ABB (2) Theben (3) Zenino (4) EnOcean

Figure 11.4　The current solutions for smart home

Ⅴ. **New Words**

luminance ['lumɪnəns]	n.	[光][电子] 亮度；[光] 发光性
actuators ['æktjuː,eitə]	n.	[电]致动器，制动器
gateway ['geɪtweɪ]	n.	网关
switchboard ['swɪtʃbɔːd]	n.	配电盘；接线总机

Ⅵ. **Phrases**

Smart homes	智慧家庭，智能家居
battery-driven	电池驱动的
maintenance-free	不需维护的，不用维护的

Ⅶ. **Abbreviations**

App　Application	应用程序
HVAC　Heating Ventilation Air Conditioning	供暖通风与空气调节
OEMs　Original Equipment Manufacturers	原始设备制造商

Passage C Smart Home System for Elderly

A smart sensor system is one with an intelligent, pervasive information and communications technology. Statics indicate that the population that is 60 years or older throughout the world is steadily on the raise. It is estimated that by 2050 this particular group will have globally increased by over 50%. This can cause an immense stress on the scarce resources available to care for the elderly. Therefore the importance of enabling the elderly to live in their own home as long as possible is crucial. And the smart home monitoring has the potential to be a cost-effective alternative, capable of providing quality health care and monitoring households for elderly. Some of the key challenges faced by wireless sensor based home monitoring are system design, communication, reliability, and privacy are discussed in this paper. The smart home monitoring discussed here, with their intelligent technological designs allow older subjects to live autonomously in a comfortable and secure environment.

Smart homes can provide the elderly with many different types of emergency assistance systems, security features, fall prevention, automated timers, and alerts. These systems allow for the individual to feel secure in their homes knowing that help is only minutes away. Moreover, smart home systems will make it possible for family members to monitor their loved ones from anywhere with an internet connection.

Typically, a smart home for elderly or disabled has the capabilities of:

- monitoring the activities of the householder and the living environment to ensure the safety of residents,
- detecting the physiological and mental condition of the householder in order to maintain the health and wellness in addition to safety,
- automating tasks that a householder is unable to perform,
- alerting the householder of potentially dangerous activities and preventing the householder from dangerous activities,
- alerting informal caregivers (family members), formal caregivers (nurses, doctors or superintendants) or first responders if the householder is in difficulties (through a linkage with a local community service scheme),
- facilitating in the rehabilitation of householders (by using auditory and visual prompts),
- linking them to the families and communities through audio-visual units (speakers, monitors, display devices, TV, etc.)

Many techniques and systems have been developed for smart home monitoring. Smart home monitoring systems with huge number of sensors are able to monitor elderly subjects. Measuring the usage of few household appliances by wireless sensors, a person's daily habits can be recognized. For this purpose smart, simple wireless sensor devices based system can be used in a house-hold of an elderly person. This intelligent wireless sensors based system will not only detect the usage pattern of the daily appliances, it will have capabilities to collate the data and flag out anomalies.

When it comes to implementation of these sensors the use and the way of implementation drastically change from one application to other application. For monitoring an old aged individual who is living independently can be done by distinguishing his/her abnormal activity from normal activity.

If there is any major deviation in the routine profile of the user, the system would trigger an alarm situation, in this case a SMS or text message could be sent to a caregiver or family member using a GSM modem. This will not only enable the smart home monitoring system to be operated remotely but also to take appropriate action immediately in case of dire need. The system also has an option for "Panic Button", wherein if the user needs an urgent help he/she can activate the alarm right away. The wireless sensors which are used in this system are centralized in structure and distributed around the house. The data acquired by the sensors are sent to a base station wirelessly via standard RF communication such as ZigBee. The core components in the system include Sensor Units, basestation, a PC which running on a smart sensor supervisory program and a cellular (GSM) modem.

Ⅰ. New Words

immense [ɪˈmens]	adj. 极大的，巨大的；浩瀚的，无边际的
autonomously [ɔːˈtɒnəməslɪ]	adv. 自律地；自治地
capabilities [ˌkeɪpəˈbɪlɪtiːz]	n. 容量；能力
anomalies [əˈnɒməlɪz]	n. 异常现象；异常，反常，不规则
deviation [ˌdiːviˈeɪʃn]	n. 背离，偏离；离经叛道的行为
centralized [sentrəlaɪzd]	adj. 集中的，中央集权的

Ⅱ. Phrases

emergency assistance systems	紧急救援系统
secure environment	安全环境
automated timers	自动计时器
audio-visual units	视听单位

Exercises

Translate the following sentences into Chinese or English.

1. Different kinds of system components including hardware elements, software algorithms, network connections, and sensors are required to cooperate with each other to provide various services in smart home.

2. The information that a smart home collects might feel like a weapon to a teenager who gets caught sneaking in after a late-night party.

3. 无论你是在工作还是在度假，智能家居将提醒你家中正在发生什么事情，而且在紧急情况下，安全系统会起很大的作用。

4. 如果一开始就考虑更大的系统，就要精心设计智能家居，尤其要考虑以后重新布线和改造。

参考译文 Passage A 智能家居概述

I. 智能家居简介

智能家居可以被看作是一个环境,计算和通信技术运用人工智能技术来操作和控制我们的物理家居系统。在智能家居中,居民进行日常工作时,使用各项传感器器件。智能家居是一个进行自动化健康监测和评估的理想环境,使用此装置,居民的生活方式可无拘无束。

智能家居可为用户测量家庭环境(如湿度、温度、亮度等),操纵家用空调(加热、通风和空气调节),并以最少的用户干预控制其状态。

智能家居系统包括各个子系统及设备如家庭自动化、闭路电视、火灾报警、音频控制和信息娱乐系统、能源管理应用等。智能家居系统的组成部分如图11.1所示。

II. 智能家居的发展史

1898年,尼古拉·特斯拉获得了远程控制船只和车辆的专利。虽然许多家电已存在几个世纪了,但随着电力分配的引入,独立式的电力或天然气器具在1990年才成为可能。20世纪初,电器和燃气器具包括洗衣机、热水器、冰箱和缝纫机。"二战"后经济扩张,洗碗机和烘干机在国内的使用提供了方便性,它们都成为可支配收入增加转变的一部分。

第一个微处理器或出现在20世纪70年代初,应用于嵌入式系统的计算器和微型计算机。1969年,霍尼韦尔厨房计算机在内曼·马库斯奢侈品专卖店售价为10000美元(在2013年为63730美元),重量超过100磅(约45公斤),该计算机宣称可存储菜谱。用户需完成为期两周的课程学习如何编程使用该装置,使用拨动开关输入和二进制光输出,以阅读或进入这些食谱选项。它有一个内置切割板和一些内置食谱。事实证明,这种霍尼韦尔厨房计算机一台都没有卖出去。

1975年,出现了第一代通用家庭自动化网络技术X10,它是电子设备的一种通信协议,主要使用电力传输线路来传输信号和控制,其中的信号包括简短的无线电频数字数据及保留最广泛的可用度。尽管存在更高带宽的替代品,X10的新组件便宜并且可随处买到,这使X10很受欢迎,全球家庭环境使用量达到几百万。到1978年,X10产品包括16条信道命令控制台,一个灯模块和一个设备模块。不久,墙壁开关模块和第一个X10的定时器也出现了。

据ABI研究机构称,到2012年美国安装了150万家居自动化系统。

III. 智能家居的系统架构

智能家居的系统架构必须满足这些要求:测量家居条件、处理仪表数据和监控家用电器。建议的方法是用微控制器的传感器测量家居环境,用带微控制器功能的执行器来监测家用电器前端。用云计算中的PaaS(平台即服务)和SaaS(软件即服务)来处理后端的数据。图11.2说明了智能家居系统的体系架构。它由以下主要部分组成:

- 带微控制的传感器:测量家居状态;微控制器解释并处理仪表数据。
- 带微控制的执行器:接收由微控制器转换的命令以执行某些操作。该命令是基于微控制器和云服务之间的相互作用发出的。
- 数据库/数据存储:从带微处理器的传感器和云服务存储数据,进行数据分析和可视化,并作为命令队列发送到执行器。
- 服务器/前端和后端之间的API层:有利于处理从传感器接收到的数据和数据库中存储的数

据。它可从网络应用程序客户端接收命令来控制执行器，也可存储数据库中的命令。执行器提出请求通过服务器使用的数据库中的命令。
- 作为云服务的Web应用：能够测量和可视化传感器数据，并使用移动设备（如智能手机）控制设备。

IV. 智能家居技术概述

关于智能家居的各种项目、科研实验和工业陈列在世界各地展示；尽管有些智能家居有特色，他们也有许多相似之处。根据不同的研究小组，智能家居项目的目标可从证明特定某种技术创新，收集可用性信息，验证测试结果向展示最新的商业技术转变。

在生活质量方面，大多数可用的智能家居技术可以分为以下三大类。

1. 安防系统
- 使独居生活成为可能的技术，包括跌倒探测器、个人应急反应系统（PERS）和药物管理系统；
- 能够在人们的生活环境中提供服务派遣和传递的技术；
- 使生活环境更加安全的技术。

2. 健康监控
- 控制慢性疾病（如心血管疾病、哮喘、糖尿病、慢性阻塞性肺病和痴呆等）的技术；
- 用于药物管理和二级防预的技术，以及支持与病患对话进行实时交互的远距照护技术；
- 全面管理不同行业消费者的个人健康记录；
- 通过加强互联性和不同信息系统之间的信息交换能力，连接和集成各种不同的技术；
- 用于形成以消费者为中心的健康及福利技术。

3. 社会联通系统
- 用于帮助老人寻求社交圈，以此扩宽社交网的技术；
- 通过传感器和其他技术与社交相互作用，提供反馈以帮助老人认定社交互动效率，更好地控制他们自己的生活。

参考答案

1. 不同种类的系统组件包括硬件、软件算法、网络连接和传感器，这些组件要相互配合，以向智能家居提供各种服务。

2. 智能家居收集的信息可能像一个工具，能够抓住一个在聚会后深夜偷溜回家的十几岁的年轻人。

3. Whether you are at work or on vacation, the smart home will alert you to what's going on, and security systems can be built to provide an immense amount of help in an emergency.

4. If you want to start with a bigger system, it is a good idea to design carefully how the home will work, particularly if rewiring or renovation will be required.

英文简历写作技巧

撰写好英文个人简历是非常重要的，也是一项艰巨的任务。英文简历结构包括基本资料、教育背景、工作经历和特长等部分。

1. Personal Information 个人基本资料

英文简历常在简历的开头便给出自己的姓名及基本信息，而不用像有些中文简历一样，另外再列表格进行描述。基本信息一般包括姓名（加粗以醒目）、住址、电话及邮箱。注意英文地址的书写形式，地址顺序从小到大。个人信息常用词汇：

- alias 别名
- native place 籍贯
- province 省
- city 市
- autonomous region 自治区
- county 县
- nationality 民族，国籍
- current address 目前地址
- permanent address 永久地址
- date of availability 可到职时间

范例：

Name: ***

Gender: Female

Birthplace: Handan, Hebei Province   Age: 24

Address: CuiZhuang Village, Fengfeng Mining Area, Handan, Hebei Province

Postal Code: 056200

Phone: 138-8888-8888

E-mail: ***@163.com

2. Education 教育背景

最近的学历要放在最前面，时间要倒序。可只写大学期间的经历，不用涉及中学、小学。毕业院校及其所在地是必须要写。学位本科用B.S，专业名称查自己学校官网上的表述。如依然在读，用Candidate for开头比较严谨；如果已经毕业，可以把学历名称放在最前。

教育背景常用词汇：

- education 学历
- educational history 学历
- educational background 教育程度
- curriculum 课程

- major 主修
- minor 副修
- courses taken 所学课程
- educational background 教育背景/程度
- social practice 社会实践
- part-time jobs 业余工作
- extracurricular activities 课外活动
- scholarship 奖学金
- excellent League member 优秀团员
- excellent leader 优秀干部
- student council 学生会
- semester 学期

3. Experience 工作经验

注意时态正确，如果是描述现在的职务使用一般现在时，过去的则使用过去时态。尽量使用主动语态的动词来描述你的工作经历。例如 "organized" "researched" "designed" "implemented"。不要出现 I was 或 I did 的句式，比如你做了一份暑期工，就说：A summer holiday job as a waiter。不要说 "I was a waiter." 或者 "I successfully managed a restaurant."对于英式、港式简历，常出现 "Reference available upon request（如需证明，可提供见证人）"。

4. Skills 技术

语言分几个层次。"Native speaker of" 指母语；从严谨的角度讲，"Fluent in" 显得更流利。关于计算机技能，可写会使用或者熟练使用的软件名，不妨用 "Frequent user of"。资格证书，如CPA、TOEFL、GRE和GMAT等。范例：

- Language
  - English   CET6
  - Spanish   Basic
- Computer skills
  - Word, Excel, Power Point   Basic Knowledge

5. 格式注意事项

（1）逗号、句号、分号、冒号、感叹号、问号和它们后面的字符之间必须留一个半角空格。

错误：Hi,I'm an entrepreneur.Can you make this app for me for free?Because,you know,I have this brilliant idea but I don't know how to code.

正确：Hi, I'm an entrepreneur. Can you make this app for me for free? Because, you know, I have this brilliant idea but I don't know how to code.

（2）逗号、句号、分号、冒号、感叹号、问号和它们前面的字符之间不可留任何空格。

错误：I think our current version is too skeuomorphic , why don't we , uh , make it flatter so that it looks more metro-ish?

正确：I think our current version is too skeuomorphic, why don't we, uh, make it flatter so that it looks more metro-ish?

（3）将长串数字分开写，降低辨识难度。

错误：1234567890

正确：123 456 7890或123-456-7890

（4）书名、杂志名、电影名、音乐专辑名、英文中出现的外文（包括拼音，人名除外）用斜体。

（5）括号弧线外要留空格，弧线内不要。

错误：Byte Press(née Tangcha) is the best Chinese-language eBook app. Period.

正确：Byte Press (née Tangcha) is the best Chinese-language eBook app. Period.